AF845829

DIVERSITY IN
HIGHER EDUCATION

Volume 1 • 1997

MENTORING AND DIVERSITY
IN HIGHER EDUCATION

DIVERSITY IN HIGHER EDUCATION

MENTORING AND DIVERSITY IN HIGHER EDUCATION

Editor: HENRY T. FRIERSON, JR.
School of Education
University of North Carolina, Chapel Hill

VOLUME 1 • 1997

United Kingdom – North America – Japan
India – Malaysia – China

Emerald Group Publishing Limited
Howard House, Wagon Lane, Bingley BD16 1WA, UK

Copyright © 1997 Emerald Group Publishing Limited

Reprints and permission service
Contact: booksandseries@emeraldinsight.com

No part of this book may be reproduced, stored in a retrieval system, transmitted in any form or by any means electronic, mechanical, photocopying, recording or otherwise without either the prior written permission of the publisher or a licence permitting restricted copying issued in the UK by The Copyright Licensing Agency and in the USA by The Copyright Clearance Center. No responsibility is accepted for the accuracy of information contained in the text, illustrations or advertisements. The opinions expressed in these chapters are not necessarily those of the Editor or the publisher.

British Library Cataloguing in Publication Data
A catalogue record for this book is available from the British Library

ISBN: 978-0-76-230086-0
ISSN: 1479-3644 (Series)

Transferred to digital printing 2005

INVESTOR IN PEOPLE

CONTENTS

LIST OF CONTRIBUTORS

Cheryl A. Boyce	National Institute of Mental Health College
Jennifer Boyce	College of Education Arizona States University
Marcia E. Canton	Applied Sciences and Arts San Jose State University
Henry T. Frierson, Jr.	School of Education University of North Carolina, Chapel Hill
Gloria Harbin	School of Education University of North Carolina, Chapel Hill
Marilyn J. Haring	School of Education Purdue University
Charles C. Healy	Graduate School Education and Information Studies UCLA
Stafford L. Hood	College of Education Arizona State University
Freeman A. Hrabowski, III	President University of Maryland, Baltimore County
David P. James	Educational Development and Degree/Extension Center Prince George's Community College

William Malloy — School of Education, University of North Carolina, Chapel Hill

Kenneth I. Maton — Department of Psychology, University of Maryland, Baltimore County

Theresa McCormick — College of Education, Iowa State University

William McHenry — Assistant Commissioner for Academic Affairs for the State of Mississippi

Olga M. Welch — School of Education, University of Tennessee

Reginald Wilson — Senior Scholar, American Council on Education

Charles Woolston — Vice Provost, University of Maryland, Baltimore County

Morris Zelditch, Jr. — Department of Sociology, Stanford University

PREFACE

It is the elusive quality of mentoring which began a five-year, cross-disciplinary conversation at American Educational Research Association meetings about mentoring in higher education that spurred the development of volumes 1 and 2 of this series. The conversation has ranged from descriptions of mentoring relationships in community college settings, to issues related to cross-race and cross-cultural mentoring, to mentoring for career advancement across organizational contexts to the phenomenom of "negative mentoring." In each case participants have sought to present their own perspectives or research on mentoring in the context of an ongoing dialogue, revealing that while mentoring relationships share some commonalities, they also retain differences that underscore their complexity. This first volume presents models and approaches of mentoring that have been employed with the purposes of increasing in the diversity in higher education ranging from the ranks of students to that of faculty.

In the first section Charles Healy provides a cogent and operational definition and related discussions of mentoring in the academic setting. He points out the concept of reciprocity as a key ingredient in the evolving relationship between mentor and protégé in the academic milieu. Morris Zelditch then details the wide and various roles that a mentor or mentors can play in the development and support of graduate students.

The second section deals with various models or approaches to mentoring in institutions of higher education. Olga Welch examines what makes an effective approach to mentoring. Marilyn Haring presents and discusses the concept of networking mentoring which is an approach quite different from the conventional view of the mentoring process. In networking mentoring, in addition to the more typical mentor-protégé relationships between faculty and student, mentoring relationships are formed among the students themselves and serve as a foundation for support. Marcia Canton and David James present overviews of successful mentoring approaches employed at two institutions that are different in scope, San Jose State University and Prince George's Community College. Cheryl Boyce discusses the notion of involving African-American graduate students in the mentoring of African-American undergraduate students to enhance their prospects of seeking to enroll in graduate programs. Freeman Hrabowski, Kenneth Maton, and Charles Woolston write about the highly successful Meyerhoff Program at the University of Maryland-Baltimore County and the important role mentoring plays in that program in producing prepared and academically talented students in the sciences. William McHenry discusses the importance of mentoring in a National Science Foundation-sponsored nationwide enrichment and research training program for minority students that has had the effect of encouraging a number of students to pursue graduate studies in the sciences. Stafford Hood and Jennifer Boyce presents a rather unique approach in their discussions of the involvement of professional associations in the development of structured mentoring programs for student members. Significantly, those association-based mentoring programs would be aimed at serving individuals of color and would further serve to increase their representation in the represented professions. William Malloy and Gloria Harbin describes the importance of mentoring in a postdoctoral program for special education faculty from historically black colleges and universities who work with research faculty from the University of North Carolina.

The two chapters by Reginald Wilson and Theresa McCormick, respectively, examine mentoring from an entirely different perspective from which we are commonly accustomed. Both authors look at the potential detrimental effects of what are called mentoring situations. Wilson describes the concept of negative mentoring which may be encountered by students of color. McCormick, on the other hand, points outs the pitfalls of mentoring that may befall individuals of color and others who she describes as on the margins in academic environments. Both authors provide a sobering look at what may pose as mentoring and warn the reader against the existence of negative phenomena that could be falsely associated with the mentoring process.

Volume 1 of this series presents concepts, models, and programs that address mentoring as it relates to higher education. It provides the reader with a number of views and approaches. The views may provide readers with different perspectives of mentoring and the approaches may give readers ideas from which they can launch mentoring programs or incorporate in their actual mentoring of individuals, or still, provide individuals with criteria to assess their own experiences as either mentors or protégés. Those conversations at AERA over the last five years have proven to be quite exciting and stimulating as witnessed by the contents of these two volumes with this volume leading to an expansion of those discussions.

INTRODUCTION

Henry T. Frierson, Jr.

It has been remarked that volumes 1 and 2 of this series are timely, for mentoring has become an increasingly popular concept. Since the late 1970s published papers on the mentoring process have begun to appear rather frequently and the establishment of formal mentoring programs have become common (Jacobi, 1991). In higher education there have been a number of definitions and meanings ascribed to the mentoring concept (Merriam, Thomas, & Zeph, 1987). Healy and Welchert (1990) and Jacobi (1991) pointed to these disparate definitions and asserted that the lack of a single comprehensive and functional definition hampers research on mentoring. In efforts to reduce what they described as confusion and generate a functional and comprehensive definition, Healy and Welchert (1990) examined the mentoring process from a developmental and contextual perspective. They maintained that such a perspective articulates the essence of mentoring and is able to accommodate the observations from studies of the mentoring process. Subsequently, they defined mentoring as "...a dynamic, reciprocal relationship in a work environment between an advanced career incumbent (mentor) and a beginner (protégé) aimed at promoting the career development of both." (See chapter 2 for a full discussion of the developmental and contextual perspective of mentoring.)

Diversity in Higher Education, Volume 1, pages 1-5.
Copyright © 1997 by JAI Press Inc.
All rights of reproduction in any form reserved.
ISBN: 0-7623-0086-8

If mentoring is important for students, particularly for those who aspire to become researchers and scholars, a foundation for the mentoring relationship must first be established. Following a Vygotskian-oriented view, before a true mentoring relationship is established, students and their prospective mentors must advance through what is in effect a pre-mentoring phase. Such a phase is aligned with the first two levels of transformation in the mentoring process described by neo-Vygotskians such as Wertsch (1985). According to Wertsch and further described by Valadez (1990), the first level is characterized by difficulty in communication because of differences in knowledge between in student and teacher. At the second level the student participates more successfully although communication problems still exist. At the third level the student can make inferences to interpret the teacher's directions. At the fourth level the student does not rely on the teacher for direction and is responsible for completing assigned tasks. The fourth level is where a true mentor-protégé relationship occurs. Because the foundation for a mentoring relationship occurs particularly during the first two levels of transformation, it follows that the more positive the student's perception of the teacher and their relationship, the more effective the student can progress through the first two stages of transformation and subsequently enter the third and fourth levels.

Mentoring in graduate education has been described as a process that provides individuals with support and protection during their graduate training and serves as additional support once they become professionals (Frierson, 1990). Cusanovich and Gilliland (1991) stated that mentoring is considered so important in graduate education that not only is it an essential part of graduate education, it may be the heart of graduate education. Blackwell (1983, 1984), however, maintained that African-American graduate students do not commonly experience the mentoring process. He reported that only one in eight black Ph.D. recipients had the benefit of a true mentor during graduate school.

Significantly, the number of underrepresented minorities receiving Ph.D. degrees and entering academe continues to remain quite small (Simmons & Thurgood, 1995; Wilson & Carter, 1995). To increase the number of students of color enrolled in graduate programs—and subsequently, faculty ranks—outreach efforts are needed to encourage African-American, Latino, and Native American students to pursue doctoral studies. A suggested approach was to provide a greater number of those students with research experiences as undergraduates (Frierson, 1981). Starting particularly in the late 1980s there was a number of minority student-oriented summer research programs initiated at research universities (Tyler, 1993). Objectives of these programs are often to give students in-depth research experiences, heighten their attitudes toward research, and encourage them to pursue graduate studies. These programs are frequently termed research mentoring programs. At most universities where these programs are, or were, administered, the proportion of minority faculty is small (Carter & Wilson, 1995). Presumably, therefore, only a small percentage of the participating students work with minority faculty preceptors.

To reiterate the views of Michael Cusanovich and Martha Gilliland (1991), who, at the time were graduate deans at the University of Arizona, considered mentoring so important .in graduate education that it may be the heart of graduate education. If what Cusanovich and Gilliland state is true, then according to Blackwell and others, African-American and other graduate students of color may be receiving graduate educations that are wretchedly unequal. Many point to the importance of students' cultural and ethnic background concerning the nature of the mentoring relationships which they can establish. In truth, background may also determine whether they will actually have a mentoring relationship.

Here it is suggested that the more positive the students' perception of the teacher and the relationship between the two, the faster the student will reach the latter stages of the neo-Vygotskians' second level of transformation and subsequently the third and fourth levels. This also suggests that the foundation for a mentoring relationship may or, as importantly, may not be established. The pre-mentoring concept is used here because the students observed in this study all started in a preceptor-student relationship with university scholars. Similarly, for graduate students such relationships may or may not evolve into a mentor-protégé relationship. This situation is essentially true, for all intent, in formal or contrived mentoring programs.

The issues addressed in this volume are significant because of the increased number of formal mentoring programs and the question of what type of relationship can be expected for students of color under these conditions? Another question is what associated effects will the preceptors' race and gender have on the perceptions and attitudes of students of color who are involved in such programs?

In discussing the lack of minority faculty and their availability as mentors for minority students, the question often arises as to whether race or ethnicity make a difference concerning who mentors ethnic or racial minority students. With the increased number of formal "mentoring" programs particularly those associated with programs for individuals of color, the implications of such issues are significant. A related question is what types of student-faculty relationships are likely to be expected for those students in formal programs? In addition to race, faculty gender may also be a significant factor. Thus, another key question is what associated effects will faculty members' race and/or gender have on the perceptions and attitudes of students of color, in this case African Americans, participating in research "mentoring" programs?

It is clearly acknowledged that mentoring is critical for a comprehensive and full, graduate education, and many see its value at the undergraduate level as well. Moreover, mentoring can have considerable consequence in one's success as an untenured faculty member. Consequently, there have been numerous efforts to formalize the mentoring process, as pointed out in this volume, that range from a focus on undergraduates to faculty. Clearly a question that could be asked of the organizers of formal programs would be, what are the ramifications for students not of the same culture, gender, or ethnicity of those assigned as their "mentors?" At most major research universities where highly recognized graduate programs exist, African Americans and other students of color often find themselves in situations where there are few, if any, faculty who are of the same culture or ethnicity. Conventional wisdom supports the notion of the need for greater numbers of faculty of color to work with racial and ethnic minority students and further to broaden the opportunity of students from the general population to interact with and have mentors from backgrounds different from themselves. This in itself would prove to promote diversity in higher education. But the main concern of volumes 1 and 2 of this series is the mentoring of individuals of color and paving an access to graduate programs and faculty positions in higher education.

REFERENCES

Blackwell, J.E. (1983). *Networking and mentoring: A study of cross-generational experiences of Blacks in graduate and professional schools.* Atlanta: The Southern Educational Foundation.

Blackwell, J.E. (1984, October). *Increasing access and retention of minority students in graduate and professional schools.* Paper presented at the Educational Testing Service's Invitational Conference on Educational Standards, Testing, and Access in New York.

Cusanovich, M., & Gilliland, M. (1991). Mentoring: The faculty-graduate student relationship. *Council on Graduate School Communicator, 24,* 1-2.

Frierson, H.T. (1981). Minority participation in R&D: Developing an undergraduate feeder system. *Journal of Negro Education, 50,* 401-406.

Frierson, H.T. (1990, March). The situation of black educational researchers: Continuation of a crisis. *Educational Researcher, 19,* 12-17.

Healy, C.C., & Welchert, A.J. (1990, November). Mentoring relations: A definition to advance research and practice. *Educational Researcher, 19,* 17-21.

Jacobi, M. (1991). Mentoring and undergraduate academic success: A literature review. *Review of Educational Research, 61,* 505-532.

Merriam, S.B., Thomas, T.K., & Zeph, C.P. (1987). Mentoring in higher education: what we know now. *The Review of Higher Education, 11,* 199-210.

Simmons, R.O., & Thurgood, D.H. (1995). *Summary report 1994: Doctorate recipients from United States universities.* Washington, DC: National Academy Press.

Tyler, M.D. (1993). *Enhancing the minority presence in graduation education V: Summer research opportunity programs.* Washington, DC: Council of Graduate Schools.

Valadez, J.R. (1990). Case studies of a mentorship program: A neo-Vygotskian perspective. Doctoral Dissertation.

Wertsch, J.V. (1985). *Culture, communication, and cognition: Vygotskian perspectives.* New York: Cambridge University Press.

Wilson, R., & Carter, D. (1995). *Minorities in higher education: Fourteenth annual status report.* Washington, DC: American Council on Education.

PART I

CONCEPTS OF MENTORING

AN OPERATIONAL DEFINITION OF MENTORING

Charles C. Healy

In an effort to revitalize our nation's competitive vigor, schools, universities, businesses, and government have co-opted a strategy of the ancient Greeks: mentoring. As Odysseus entrusted his son and posterity to a sage elder, moderns have recognized a resource in seasoned educators and seek to exploit their wisdom for future generations of teachers, scholars, professionals, and managers. A movement begun in the 1970s, as many corporations and government agencies launched mentoring programs, continued unabated in the 1980s as colleges, universities, school districts, and states established programs to enhance the quality of faculty and administrators through mentoring.

The desire to enlist mentors in optimizing career development and, more recently, in promoting excellence in education, has inspired a flurry of research and development projects on mentoring (Gray & Gray, 1986). Unfortunately, the seeds of empirical study have been cast too broadly to yield a harvest of cumulative knowledge given that inconsistent, idiosyncratic definitions of mentoring are employed which frequently lack grounding in theory (Bogat & Rednar, 1985; Carden, 1990; Merriam, 1983). Absence of a definitional consensus

Diversity in Higher Education, Volume 1, pages 9-22.
Copyright © 1997 by JAI Press Inc.
All rights of reproduction in any form reserved.
ISBN: 0-7623-0086-8

is stymieing efforts to synthesize empirical findings into a coherent body of knowledge and to identify important unanswered questions.

We are not the first to note that mentoring is referred to in widely disparate terms that impede investigation: "One problem in the mentoring literature is the lack of any one comprehensive, yet functional, definition" (Bogat & Rednar, 1985). However, a frequent response to the definitional conundrum has been a nod in its direction while proceeding with an admittedly inadequate formulation. Our purpose, consequently, is to advance a definition of mentoring relationships, derived from current thinking in developmental-contextual theory, that is both functional and comprehensive. By functional we mean a definition that articulates the essence of mentoring so that it is distinguished from other superior/subordinate interactions, bridges the hiatus between formalized and classical mentoring, and implies corollaries that highlight significant unanswered questions. By comprehensive we mean a definition that is able to accommodate the observations from studies of mentoring relationships.

Our definition incorporates the developmental-contextual notions that stage denotes a qualitatively distinct level of organization and that context both influences the organism's development and is changed by it (Vondracek, Lerner, & Schulenberg, 1986). It expands Levinson's (1978) influential developmental definition by postulating that an organism's transformation depends as much upon the dynamic potentials of its context as upon its own changing capacities. This formulation conceptualizes development as a nexus of dynamic, bidirectional organism-context interactions with probabilistic outcomes.

Within this framework we consider mentoring to be a dynamic, reciprocal relationship in a work environment between an advanced career incumbent (mentor) and a beginner (protégé) aimed at promoting the career development of both. For the protégé the object of the mentoring relationship is achievement of an identity transformation, a movement from the status of understudy to that of self-directing colleague. For the mentor the relationship is a vehicle for achieving midlife "generativity" (Erikson, 1963), meaning a transcendence of stagnating self-preoccupation via exercise "of an instinctual drive to create and care for new life..." (quoted in Yamamoto, 1988, p. 186).

Two elements of the definition are fundamental and distinguish mentoring, whether formal or informal, from other superior/subordinate helping relationships such as teaching and supervising: reciprocity between mentor-protégé and accomplishment of an identity transformation by each party. With respect to reciprocity, many discussions of mentoring acknowledge that benefits may accrue to the mentor, but potential for mentor development is typically regarded as a serendipitous by-product rather than as an integral constituent of the relationship. The proposed definition posits mutual exchange as a sine qua non of mentoring.

Contrasting mentoring with supervising and teaching clarifies the reciprocity in mentoring. The mentor, but not the supervisor per se, engages in interactions vis-à-vis the protégé aimed at passing on his or her professional legacy—a manifestation of generativity. Mentors transmit a complex legacy of professional acumen that reflects their own unique ability to identify salient issues and heuristics in the work environment. They cultivate qualitative changes in the protégé's approach to tasks rather than his or her immediate productivity.

Mentors, but not the teacher qua teacher, invite protégés to partake of the wisdom and style which have enabled them to excel professionally. These professional troves are not concepts and methods from textbooks but approaches with which the mentor applies the knowledge of the craft. Zuckerman (1977) noted that as the relation blooms, protégés integrate aspects of this professional approach into their repertoire. Their initially unequal relationship with the mentor becomes reciprocal as their practice incorporates and thereby perpetuates essential elements of their mentor's professional legacy. In adapting the art of the mentor to new challenges, protégés become living transmitters of their mentors' artistry. Although mentors' collaboration in this adaptation may assist them in upgrading their professional expertise, the distinctive achievements of a mentor appear to be selfless transmission of one's professional legacy to posterity by empowering a protégé in its rendition and the intrinsic satisfaction of such a contribution.

Given our definitional emphasis on the contextual nature of mentoring, it follows that the evolution of the relationship and the developmental trajectories of its participants are influenced by many interacting variables, including mentor/protégé personalities, their shared and distinctive histories, and their mutual and separate contexts. Clearly, mentoring dynamics do not depend only on events circumscribed by the relationship. For instance, if a mentor's guidance on a particular instructional strategy improves student evaluations of the novice teacher, the protégé is likely to become more self-revealing and imitative of the mentor. The broader context, however, such as the school's culture and family demands also affect the protégé's satisfaction with such experimentation and propensity for self-revelation and imitation.

The proposed definition is applicable not only to those "spontaneous" mentorships that arise by mutual consent between parties but also to mentorships established via formal programs. Moreover, efforts to institute mentorships without reference to a definition that captures their essence may trivialize them. Little (1990), for instance, reports that many teacher mentoring programs are falling short of expectation because they were initiated without sufficient consideration of either their structure and objectives, or of the institutional conditions necessary for them to succeed in educational settings. More critical of institutional mentoring programs, Yamamoto (1988) believed that much of what passes for mentoring, for example, formalized assistance in networking and coaching in professional skills, degrades "yet another human phenomenon of profundity...by a misguided attempt at popularization and standardization." Gehrke (1988), too, warned that mentorships are debased when they are viewed as self-promoting business arrangements rather than as transforming gift exchanges. Hardcastle (1988) likewise suggested that the core qualities of "significant" mentorships are absent from formal affiliations labeled as mentoring.

What emerges from the literature is a polarized view of mentoring. On the one hand there is "classical" mentoring worthy of the name—it is dynamic, occurs spontaneously between two people of goodwill and commitment, is long term, multifaceted, and potentially profound in impact. On the other hand are assigned, short-term, cost-effective arrangements of limited significance that have, in the minds of some, sullied and usurped the title "mentoring." Formalized mentoring (read inferior) lends itself to quick, albeit myriad operational definitions and critiques, while true mentoring has an ineffable quality that ought not be violated by crassly quantitative methods (Gehrke preferred "affectionate research").

However, thinking that pits "true" mentoring against "imitation" mentoring begets an unproductive state of affairs. For one thing, the conclusion that deliberate attempts to foster mentoring are doomed to yield limited results is premature because formalized mentoring programs are relatively recent phenomena and there has been little time to hone and evaluate them. For another, the essence of mentoring has not been explicated sufficiently to distinguish institutional mentoring from other staff development programs. Thus, the suggestion that intentional mentoring debases a human phenomenon of profundity is a hypothesis to be tested, not a truism to be affirmed.

Already there is limited evidence that the essence of mentoring as advanced by the developmental-contextual definition—reciprocity and potential for qualitative transformation—may be tapped by a formal program. An evaluation of the California Mentor Teacher Program (Ruskus, 1988) indicated that some participating mentors did reap career benefits far beyond the nominal $4,000 annual stipend they received. Mentors interviewed cited collegiality, a positive sense of efficacy, the opportunity "to have a broader impact than they would have as regular teachers," and the exercise of leadership, as benefits. One mentor (p. 219) felt that the program facilitated her creativity, another (p. 219) that it allowed him leverage to have an impact on adults. A master teacher (p. 220) said the mentor role allows mentors to reinforce and refine their own philosophies and is thus professionally enriching. Several mentors (p. 221) said the program allowed them to grow professionally in ways not possible without it; others (p. 222) mentioned that they had more control over their own destinies. The perceptions of these mentors indicate that they not only gave assistance to protégés but received benefits significant to their own growth. These examples suggest that formal, appropriately structured programs may promote the developmental-contextual hallmarks of reciprocity and qualitative transformation and need not degrade a profound human relationship.

SAMPLING IMPROVEMENTS

A major methodological flaw in many of the studies reviewed for this paper was a tautological definition of mentoring that produced positively biased samples. In these investigations questions to respondents elicited only the recall of relationships with positive outcomes, for example, had they an association with a senior worker instrumental in advancing their careers? Asked in this form, less successful examples of mentoring likely go unreported, and we do not learn from the failures. Unfortunately, however, failures in mentorships occur as Carden (1990) and Rogers, Holloway, and Miller (1990) have pointed out. Indeed, after reviewing many of the early studies of formalized mentorship programs for teachers, Little (1990) suggested that while long-term mentorships tend to be enhancing for participants, brief mentorships may be deleterious, weakening the protégé's teaching skills. By establishing the domain of mentoring relations a priori as those designed to promote reciprocity and further the goal of participant transformation, our definition may eliminate biased sampling and permit us to pursue the contributing parameters of mentorships.

Lacking a consistent definition of mentoring, those qualities considered desirable in a mentor have often been mistakenly advanced as criteria for the mentoring function. A case in point is "sensitivity." The sensitivity of the mentor to the protégé's individuality and tact in giving feedback refer to the mentor's *skills*, not to the *type of relationship*. These qualities cannot be invoked to categorize relationships as mentoring or non-mentoring since they emerge during the course of the association and permit only ex post facto classification. For example, when Clawson (1980) found that outstanding supervisors were more likely than their marginal counterparts to relate to subordinates in a manner characterized by such elements as mutual respect, agreement about one another's roles, consistency, and informality, he concluded that these elements distinguished effective from ineffective supervision, instead of supervision from non-supervision. In the same manner, determining whether such virtues as mentor interpersonal sensitivity are requisite for productive mentoring or whether certain relational dynamics are exclusive to mentoring (as opposed to teaching or supervising), require adoption of inclusion and exclusion criteria for mentoring associations such as proposed here.

MENTORING AND DEVELOPMENT

Our review covered published research until 1994 that labeled relations between workers as "mentoring" or some synonym such as "grooming." Studies of relations between graduate or professional school students and professors were included in our review, but those between other students and their teachers were not. We assessed whether that literature supports the contextual-developmental definition of mentoring as a reciprocal association between superior and subordinate that affects their mutual transformations. We also sought empirical evidence of particular mentoring dynamics that might account for such a transformation, and documented instances of mentor/protégé development.

PROTÉGÉS

From the developmental perspective protégés do more than become proficient in specific tasks. Their proficiency is broader and greater because it includes capacity to judge when and how to apply their skills, evaluate their efforts and in other respects function autonomously. The process of transformation reaches fruition when the protégé, vis-à-vis mentor, is no longer understudy but a peer. Protégés' new collegial status evolves through a continual refinement of their work functions as they exercise increasingly sophisticated, independent judgments (Dalton, Thompson, & Price, 1977). The imbalance between superior/subordinate shifts to an affiliation between equals as mentors and protégés alike acknowledge the culmination of this transitional period.

Many studies document protégé gains beyond those of mere task competency, giving credibility to the notion of qualitative developmental change. Zuckerman (1977) found that under current or soon-to-be Nobel Laureates, future laureates incorporated standards of work, modes of thought, and identities as elite scientists. However, substantive knowledge gain was not a product of their mentoring. Shelton (1982) reported that women completing a year-long formal mentoring program were more promotable than matched controls; Cameron and Blackburn (1981) found that both formally and informally established mentorships were effective in socializing college administrators. Dalton, Thompson, and Price (1977) observed that as protégés gained technical competence and confidence in their own judgment they no longer required close supervision; Bova and Phillips (1984) observed that protégés not only honed specific job competencies, but also acquired risk-taking and political skills; Phillips (1977) reported similar gains for protégés and discovered a bidirectional exchange in the relationship as protégés assisted mentors collegially.

Mentoring studies elucidate the probable means as well as the modes of protégé transformation. Elite scientist mentors evoked excellence by setting high standards and by illustrating the art of intuiting important questions (Zuckerman, 1977). Relatedly, Hardcastle (1988) found mentors set high standards and gave protégés unique insights into their potential. Kram (1985) and Phillips-Jones (1982) observed a number of beneficial functions performed by mentors in the corporate world such as sponsorship and exposure. Gehrke (1988) inferred that mentors awaken protégés to their potential and elicit their commitment to express it. Several of the artists and scientists who John-Steiner

(1985) interviewed described experiences consistent with Gehrke's inference. As a consequence John-Steiner proposed that mentorships transform potential to expressed creativity by providing the beginner with insights into both the overt activities of human productivity and into the more inner processes of thought. Through close collaboration the novice can learn what the mentor may not even know; how he or she formulates a question or starts a new project. Their review of mentoring studies led Rogers, Holloway, and Miller (1990) to a similar conclusion. For them the powers of mentorships lie in the intellectual and emotional energy they create and consume. Mentorships sanction mutual validation, reciprocity, a relaxed atmosphere and trust which allow risk taking and self-disclosure. Particularly important, they contend, is a mentor's self-disclosure to a protégé of "a sense of inadequacy or fear of being vulnerable (secret goals, weaknesses and worries)."

These studies support the thesis that some protégés achieve developmental advances through mentoring, yet they do not confirm Levinson's (1978) contention that mentoring relationships are indispensable avenues to transformation. Even though some professional accomplishments, for example, those mentioned by Conway (1981) for nursing, may be attainable only through a mentor's grooming, many workers have matured without mentoring (Phillips, 1977; Kram, 1985, Rogers & Holloway, 1993), and some in spite of it (Watkins, 1980; Grant & Zeichner, 1981). For instance, Rogers and Holloway discussed how peer relationships can assist beginners substantially in their transformation from novice to self-directing professionals. When protégés estimated the contribution of mentoring to their career development, moreover, many indicated that it was not one of the most important factors (Anderson & Devanna, 1981; Phillips, 1977).

MENTORS

The contextual-developmental definition maintains that mentors as well as protégés can be transformed through mentoring. Career development in professional and managerial fields occurs as workers successfully redefine their roles, assuming capacities such as specialist, troubleshooter, consultant, leader, mentor, while they simultaneously balance other life roles (Dalton, Thompson, & Price, 1977; Super, 1990). To achieve developmental transformation, advanced career incumbents must press toward higher levels of functioning to serve future generations. For example, they might integrate new advances in their fields into their existing skill repertoires, introduce their own innovations or help in managing work in the field and coordinating its functions with related fields. They might also transmit their fund of experience to followers and consultees.

Mentoring functions are conducive to precisely these kinds of advances and thus it is plausible that mentors, in the very act of guiding and promoting others, act to affect their own transformations. Indeed, evidence is amassing that mentoring potentiates *developmental change*. Blackburn, Chapman, and Cameron (1981) observed that academic mentors derived gratification from collaborating with protégés to produce new knowledge; Dalton, Thompson, and Price (1977) and Kram (1985) reported that successful mentoring was recompensed with intrinsic satisfaction, added responsibilities, and leadership recognition; and Atkinson, Casas, and Neville (1994) found that ethnic minority psychologists felt that the process of helping a protégé in professional development and the attendant increased enthusiasm for their own work through the process were the most satisfying aspects of mentoring.

Corollary #1: Readiness

The proposition that development progresses through discrete stages has several corollaries action research may explore to enhance mentoring. One corollary is that the degree of maturity that both parties bring to the relationship influences its outcome. Although researchers have not investigated protégé motivation and readiness formally, they have identified qualities that enhance mentoring which appear to reflect maturity level. For instance, individuals must often show exceptional promise to be accorded protégé status (Hall & Sandler, 1983; Kanter, 1977; Phillips, 1977). Additionally, many Nobel Laureates intentionally sought out elite mentors (Zuckerman, 1977), implying that constructive mentorship may be *secured* through proactive, systematic career planning, rather than *conferred* by another. Dalton, Thompson, and Price (1977) hypothesized that MBAs, scientists, and engineers new to organizations appear to need sufficient maturity to relate effectively to authority while simultaneously maintaining their autonomy.

The literature is far richer on the topic of mentor readiness, implying that professional maturation and status necessarily affect the quality of mentoring. It is intuitively reasonable that those best able to promote the growth of others must have first developed their own capacities. Dalton, Thompson, and Price noted that the most qualified technical, professional, and managerial mentors assumed increased responsibility for guiding the career development of others and had expanded their network of contacts to benefit their protégés. Others (e.g., Kanter, 1977; Kram, 1985; Schein, 1978) also observed that mentors must possess power and professional respect to garner opportunities for their protégés. Missiarian (1982) underscored the importance of professional influence by calling "mentor's power" the third of her three dimensions for distinguishing among mentorships. Certainly, the notion of reciprocity suggests that continual, high-level performance by mentors propels protégé development; conversely, a slackening of mentor productivity may inhibit it.

Corollary #2: The Developmental Process of a Mentoring Relationship

A second corollary is that mentorships pass through qualitatively distinct periods. This is consistent with the findings of Phillips (1977) and Kram (1985) that mentoring relations pass through periods that are distinguishable. Phillips named the stages of mentoring: mutual admiration, development, disillusionment, parting, and transformation, reflecting the developmental advances of the protégé. With Phillips's nomenclature as a point of departure, Kram (1985) studied mentoring dyads and proposed a different nomenclature: initiation, cultivation, separation, redefinition.

Both terminologies portray a similar sequence: initially both participants get a sense of one another as the mentor assists and observes the protégé. As their interaction becomes motivating and commands energy, both experience increasing mutuality and willingness to risk for one another. During this time optimum learning occurs and maximum accomplishments result. Phillips and Kram agree that this intense interaction gradually wanes as the parties distance themselves; at the separation juncture they must either redefine their relationship based on collegiality or suffer a deteriorating alliance.

Phillips and Kram have deepened our understanding of mentoring relations. Their work supports the inference that two indispensable elements promoted an optimally enhancing relationship: mentor and protégé must experience the relationship as a reciprocal venture; each party must increasingly esteem the other. Their findings can be interpreted as showing that participants need to perceive mentoring as a quid pro quo arrangement for them to invest sufficient energy to promote growth. Mentors, no less than protégés, may require assurance that their partners will give as well as receive.

Phillips and Kram underscored the necessity of evaluating the process of mentoring, from inception to resolution, to verify the transformations posited by the stage construct. They challenged researchers to discover factors in the mentorship, including its method of initiation and its parties' contexts, which advanced it to the mutuality phase. Neither insists that mentor and protégé choose each other at the outset, but both hypothesize that commitment to one another as people and appreciation of one another's professional contributions increase the possibility of transformation. Other important research questions they suggest are: What percentage of mentorships in particular environments advance to the mutuality stage? What are the outcomes of mentorships that terminate before mutuality is achieved?

Corollary #3: Mentoring and Participants' Contexts

A third corollary of the contextual-development definition is that mentorships change participants' contexts even as contexts shape mentoring. Initial mentoring transactions will likely abide the culture and organization of their host sites, but as a mentorship evolves, its transactions can diverge from and change those parameters. Therefore, the impact of a mentorship cannot be appreciated unless change in contexts is taken into account. In a school where team teaching is out of vogue, for instance, mentor and protégé are unlikely to team-teach at the start of this relation. But as mentoring succeeds, they might arrange joint efforts, and in so doing prompt their colleagues to reevaluate the team approach. Certainly the mentoring team's perceptions of contextual responses are likely to influence their subsequent transactions, so that understanding the course of mentoring requires examining interactions with the parties' contexts.

Others such as Hunt and Michael (1983) have already noted the importance of considering contextual variables in understanding mentorships and have hypothesized that the relation between mentorships and their contexts is bidirectional. And already several studies have indicated that contextual variables affect mentorships. For example, Little (1990) showed the importance of school organization and teacher's ethos on the course of teachers' mentoring, and Koberg, Boss, Chappell, and Ringer (1994) found that both the effectiveness of an employee's immediate work group and organizational rank influence the likelihood of the employee being mentored. Yet as Koberg and his colleagues (1994) pointed out, the influence of context on mentorships is understudied, and the effect of mentorships on their context appears to have been studied even less often. Since bidirectional influences are central to the contextual-developmental model, its appreciation by mentoring researchers will show the importance of studying how mentorships change their contexts.

The corollary also warns against overlooking the parties' nonwork roles. By introducing new significant others into their lives and adding new responsibilities and prerogatives to the work role, mentorships alter their participants' life space. The roles participants must balance along with mentoring are likely to affect the nature and consequences of the transactions in a mentorship, and the mentorship is likely to affect participants' immediate roles. Phillips (1977) noted that mentorships of single, opposite-sex partners occasionally ended in marriage, and Kram (1985) reported that one of the mentorships she observed terminated because of the mentors' problems with a teenage son. Disregarding the parties' life space can only blur appreciation of their mentorships.

Corollary #4: History and Age-graded Influences

A fourth corollary is that the concepts of normative history-graded influences (i.e., those specific to a cohort arising from idiosyncratic events in the generation) and age-graded influences (i.e., those stemming from bio-social maturation) illuminate factors that will affect the contribution of the mentorship to its participants and host institution. Kanter (1977) pointed out that mentorships provide "sponsored" as opposed to "contest" mobility. If they are to support our society's commitment to equal opportunity, consequently, a person's gender, ethnicity, or religion should not improve or lessen his or her chances of becoming a protégé. Yet, given the norms operative during their socialization, we would expect the historically advantaged white male to choose white male protégés, despite his endorsement of equal opportunity. Underrepresentation of women and people of color among the ranks of protégés would follow, and indeed, this has been documented in some studies (Adler, 1976; Bogat & Rednar, 1985; Hall & Sandler, 1983), but not in others (Anderson & Devanna, 1980; Busch, 1985; Cameron & Blackburn, 1981; Thomas, 1990). And in a recent survey of hospital employees, Koberg and his colleagues (1994) found that the nonwhite minorites in their sample reported higher rates of mentoring than the white respondents.

Recognition that historical factors may put mentorships in opposition to equal opportunity underscores the importance of identifying ways of countering such a possibility. Studying matters such as what supports and impedes cross-gender and cross-ethnic group mentorships and what personal and contextual benefits accrue from such relationships needs to receive as much support as studying how to promote and expand mentorships. Since the collaboration of two talented individuals, who share a strong commitment to their work but have different backgrounds, may be enhanced if they can use their different lenses to improve the way they frame and address problems, research on the latter topic may provide additional evidence of the value of mentorships.

As we move to the twenty-first century the concepts of age- and history-graded influences suggest still other aspects of people's changing experiences that are likely to affect mentorships. For example, the concepts together highlight the desirablilty of studying mentorships involving protégés who are chronologically older than their mentors. Such mentorships are becoming more common as workers in their thirties and fortiess who have established themselves in one professional occupation move into another occupation. Mentorships with such nontraditional participants will likely require new kinds of readiness and offer new developmental opportunities. The notions of age- and history-graded influences suggest it may be challenging for a recycling protégé to acknowledge the wisdom of a younger mentor and to muster the energy and enthusiasm to advance that mentor's legacy. Relatedly, guiding the professional reformation of an older person who has achieved comparable occupational status in another field may require modifications by mentors; certainly experiences of adult educators suggest this (Knox, 1978).

While posing new challenges, nontraditional mentorships will likely afford expanded developmental opportunities for both protégé and mentor. Older protégés' maturation and earned self-esteem should enable them to collaborate more fruitfully. Their interactions with their younger mentors may become more egalitarian sooner and their previous accomplishment could then enable them to more easily apply that expertise to challenges in the new profession. As a consequence their achievement of a second professional identity may be a more comfortable, even fuller expression of self than the realization of their first professional identity. The legacy of mentors, in turn, may be enriched through collaboration with a protégé whom they may more quickly regard as an equal. Such enrichment may even prompt them to refine their legacy and articulate more fully what it involves. That should allow a fuller and a more substantive portion of their professional acumen to be passed on, thereby affording them a grater sense of contribution.

CONCLUSION

A definition of mentorships grounded in contextual developmental theory can advance research and practice. The definition accommodates myriad observations about mentorships from past research. It enables mentorships to be studied from their outset and to be distinguished from other relations. As a consequence researchers can examine the full spectrum of mentorships, can learn from unsuccessful as well as successful relationships, and can test how the benefits of mentorships differ from other on-the-job associations. Contextual-developmental corollaries, moreover, illuminate important research issues such as: participant readiness, distinct stages in the relationship, the bidirectional nature of context-mentorship interactions, and history- and age-graded influences.

For those who would enter or support mentorships, the definition provides both understanding and direction. It identifies potential benefits for both parties and illuminates factors that influence the roles of the participants. It points out that mutuality increases the likelihood of participant transformation, and it identifies participant maturation and context receptivity as factors that can increase the likelihood of successful mentoring.

REFERENCES

Adler, N.E. (1976). Women students. In J. Katz & R.T. Hartnett (Eds.), *Scholars in the making: The development of graduate and professional schools* (pp. 202-203). Cambridge, MA: Ballinger Publishing.

Atkinson, D.R., Casas, A., & Neville, H. (1994). Ethnic minority psychologists: Whom the mentor and benefits they derive from the process. *Journal of Multicultural Counseling and Development, 22*, 37-48.

Anderson, C., & Devanna, M.A. (1981). Mentors: Can they help women get ahead? *Career Development Activities, 2* (2), 5-8.

Blackburn, R.T., Chapman, D.W., & Cameron, S.M. (1981). Cloning in academe: Mentorship and academic careers. *Research in Higher Education, 15* (4), 315-327.

Bogat, C.A., & Rednar, R.L. (1985). How mentoring affects the professional development of women in psychology. *Professional Psychology, Research and Practice, 16*, 851-859.

Bova, B.M., & Phillips, R.R. (1984). Mentoring as learning experience for adults. *Journal of Teacher Education, 35* (3), 16-20.

Busch, J.W. (1985). Mentoring in graduate schools of education: Mentors perceptions. *American Educational Research Journal, 22*, 257-265.

Cameron, S.W., & Blackburn, R.T. (1981). Sponsorship and academic career success. *Journal of Higher Education, 54*, 369-377.

Carden, A.D. (1990). Mentoring and adult career development. *The Counseling Psychologist, 18*, 275-299.

Clawson, J. (1980). Mentoring in managerial careers. In C.B. Derr (Ed.), *Work, family, and the career* (pp. 144-165). New York: Praeger.

Conway, M.E. (1981). The national research service rewards with emphasis on the mentor relationship. *Nursing Research, 30*, 376-378.

Dalton, G.W., Thompson, P.H., & Price, R.L. (1977, Summer). Four stages of professional careers: A new look at performance by professionals. *Organizational Dynamics*, 19-42.

Erikson, E.H. (1963). *Childhood and society* (2nd ed.). New York: Norton.

Gehrke, N. (1988). Toward a definition of mentoring. *Theory Into Practice, 27*, 190-194.

Grant, C.A., & Zeichner, K.M. (1981). Inservice support for first year teachers. *Journal of Research and Development in Education, 14*, 99-111.

Gray, W. A., & Gray, M.M. (1986). *Mentoring: A comprehensive annotated bibliography of important references*. Vancouver, B.C.: International Association for Mentoring.

Hall, R.M., & Sandler, B.R. (1983). *Academic mentoring for women students and faculty: A new look at an old way to get ahead.* Washington, DC: Association of American Colleges. Project on the Status and Education of Women.

Hardcastle, B. (1988). Spiritual connections: protégé's reflections on significant mentorships. *Theory Into Practice, 27*, 201-208.

Hunt, D.M., & Michael, C. (1983). Mentorship: A career training and development Tool. *Academy of Management Review, 8*, 475-485.

John-Steiner, V. (1985). *Notebooks of the mind: Explorations of thinking*. Albuquerque, NM: University of New Mexico Press.

Kanter, R.M. (1977). *Men and women of the corporation*. New York: Basic Books.

Knox, A.B. (1978). *Adult development and learning*. San Francisco: Josey Bass.

Koberg, C.S., Boss, R.W., Chappell, D., & Ringer, R.C. (1994). Correlates and consequences of protege mentoring in a large hospital. *Group and Organization Management, 19*, 219-239.

Kram, K.E. (1985). *Mentoring at work: Developmental relationships in organizational life.* Glenview, IL: Scott, Foresman & Co.

Levinson, D.J. (1978). *Seasons of a man's life*. New York: Alfred A. Knopf.

Little, J.W. (1990). The mentor phenomenon and the social organization of teaching. *Review of Research in Education, 16*, 297-352.

Merriam, S. (1983). Mentors and protégés: A critical review of the literature. *Adult Education Quarterly, 33*, 161-173.

Missirian, A. (1982). *The corporate connection: Why executive women need mentors to reach the top*. Englewood Cliffs, NJ: Prentice-Hall.

Perry, O.G. (1987). *Structured mentoring: Socialization of Women Community College Administrators*. Unpublished Ph.D. dissertation, The University of Southern California.

Phillips, L.L. (1977). *Mentors and protégés: A study of the career development of women managers and executives in business and industry*. Unpublished Ph.D. dissertation, University of California, Los Angeles.

Phillips-Jones, L. (1982). *Mentors and protégés*. New York: Arbor House.

Rogers, J.C., & Holloway, R.L. (1993). Professional intimacy: Somewhere between collegiality and personal intimacy. *Family Systems Medicine, 11*, 263-270.

Rogers, J.C., Holloway, R.L., & Miller, S.M. (1990). Academic mentoring and Family Medicine's research productivity. *Family Medicine, 22*, 186-190.

Ruskus, J.A. (1988). *A multi-site evaluation of the California mentor teacher program.* Unpublished Ph.D. dissertation. University of California, Los Angeles.

Schein, E.H. (1978). *Career dynamics: Matching individual and organizational needs.* Reading, MA: Addison-Wesley.

Shelton, C.K. (1982). *The relationship of mentoring and behavioral style to selected job success variables.* Ph.D. dissertation, University of Southern Illinois.

Super, D.E. (1990). A life-span, life space approach to career development. In D. Brown, L. Brooks & Associates (Eds.), *Career choice and development: Applying conemporary theories to practice*. San Francisco: Jossey-Bass Publishers.

Thomas, D.A. (1990). The impact of race on managers' experiences of developmental relationships(mentoring and sponsorship): An intraorganiztional study. *Journal of Organizational Behavior, 11*, 479-492.

Vondracek, F.W., Lerner, R.M., & Schulenberg, J.E. (1986). *Career development: A life span developmental approach.* Hillsdale, NJ: Lawrence Erlbaum Associates.

Watkins, B. (1980). *Training educational opportunity researchers: Some sobered thoughts in mentoring, some optimistic thoughts in community* (ERIC Ed.), pp. 193-322.

Yamamoto, K. (1988). To see life grow: The meaning of mentorship. *Theory Into Practice, 27*, 183-189.

Zuckerman, H. (1977). *Scientific elite: Nobel laureates in the United States.* New York: Free Press.

MENTOR ROLES IN GRADUATE STUDIES

Morris Zelditch, Jr.

INTRODUCTION

My purpose in this chapter is to explore the nature and meaning of "mentoring." Everyone nowadays seems to be talking about "mentoring." Several years ago I even found an ad for "The Clairol Mentor Program" in *Women's Day* (March 20, 1990), offering to improve the mentoring of female employees. It is reasonably well established that mentoring is important: it affects whether graduate students complete the Ph.D., how long it takes them (Turner, 1972), how productive they are, and how they are placed (Reskin, 1979). But it is also problematic. It is a problem in the first instance because it appears to be the exception rather than the rule (Friedman, 1987). Even when present, it is sometimes thought to be a threat to the independence of the student (e.g., Council of Graduate Schools, 1990, p. 23). And as the diversity of the graduate student population increases, new questions arise about matching mentors to students—does matching them require that mentors be similar to students in race, ethnicity, and gender?

My focus will center around three points. First, mentoring is an ambiguous term, involving so many different roles that in fact it would prove tedious to

Diversity in Higher Education, Volume 1, pages 23-37.
Copyright © 1997 by JAI Press Inc.
All rights of reproduction in any form reserved.
ISBN: 0-7623-0086-8

cite the whole list. But second, because of the ways in which graduate education differs from undergraduate education, mentoring is an especially important means to the ends of graduate education. Third, there is another "but": because of the complexity of mentoring, its implications for the structure of graduate education are sometimes conflicting, requiring some thought about priorities among alternative ends.

Thus, what I will discuss comes down to what mentoring is, why it is important, and its implication in graduate education.

MENTOR ROLES

Etymologically, "mentor" denotes a surrogate father (Mentor was Telemachus's advisor during the absence of Odysseus). Because Athena took the form of Mentor when she wished to advise Telemachus it carries also the connotation of godlike advisor. But in universities "mentor" has accumulated a large number of other meanings. If we regard the root meaning as that of personal advisor, it has come in university settings to mean, in the first instance, several other kinds of advisor. For example, it also means an academic advisor, that is, a faculty member who advises a student about program planning, monitors academic progress, and provides feedback on performance at each of the many hurdles that constitute the race to the degree.

Thus, the Clairol ad, sponsored by the National Women's Economic Alliance Foundation, to which I referred, defines a mentor as "someone with career experience willing to share that knowledge"—that is, an advisor. She or he helps with questions such as how and where to publish your first paper, which opportunities contribute most to professional growth, how to behave in an oral exam, what jobs to apply for, what jobs to take, and so on.

But a mentor in a graduate school is more than an advisor. In part this is because, more than undergraduate education, graduate education is concerned with creating, not just transmitting, knowledge. Because knowledge grows so rapidly, graduate education tends to focus less on what is known than on methods, on means of creating new knowledge. Methods, in turn, are essentially skills, which are best learned by doing. Indeed, much of the cognitive learning in graduate schools is not in books and lectures, it is a matter of direct experience. Some skills are routines that can be learned from books and lectures, with some not particularly individualized feedback on exams, quizzes, and exercises, such as basic statistics. But higher levels of skill are not well-codified. No book describes them, and even when they do, lectures are an inadequate means of teaching them.

In my own field—experimental sociology—much of the practice of experimentation is unwritten, for example, how the experimenter creates the authority that makes experiments with human subjects possible. And much of it lies in dress, manner, props, and speech style, learned in actual performance. But even skills, like language, that are well-codified in books, and often taught by lectures, nevertheless, are not very well taught without practice, without actually speaking the language. On the other hand, experience in itself is often not enough. Unguided direct experience, though effective enough, is a costly, uneconomical way to learn. A more economical way to learn research, for example, is by rehearsing performance guided by a more experienced researcher who provides individualized feedback, both pro and con, about everything from conceptualizing a problem to designing research to measuring variables to analyzing, writing up, and presenting results. Thus, the student is essentially a research apprentice and a mentor is not only an advisor, she or he is also a research supervisor willing to tutor apprentices.

But much experiential learning is not learned from direct instruction, it is learned by observation. Kuhn, in *The Structure of Scientific Revolutions*, has emphasized how much of science is learned from exemplars, that is, models. In his view it is by mimesis that normal science is accomplished, though he is thinking more of the product than the individuals who produce it. Mimesis is a well-studied phenomenon (see, e.g., Bandura, 1986). It is in part learning simply by observing how another does something, in part learning by observing the consequences (e.g., success or failure, reward or penalty) of what another does. That is, it is in part vicarious conditioning. One learns what works and what does not work for others. Imitation of a fruitful way of doing research may become the basis of a whole program of subsequent research.

But another kind of modeling also plays an important role in graduate education. It arises from the fact that a good deal of what is learned in graduate education is not cognitive, it is the learning of values and norms because graduate education is in part vocational training, it socializes pre-professionals into the norms and values of a profession. Some of this is explicit. One learns slogans like "publish or perish." If one is a sociologist, one is explicitly taught "value-neutrality," that is, that the profession values "objectivity," strives to be independent of political, social, or economic values and interests. But much of it is latent learning, implicit, even unconscious. Tastes in problems are often of this kind. A sociologist in the early 1990s learns that the relations of the organization to the environment or the political system of a nation to the world system are "hot" topics, but the family and criminology are not, and human sexuality is "pop" and therefore trivial. Not that any sociologist openly says that human sexuality is not worth studying, it's just an assumption that is communicated by how models act. Many of the markers of professional identity are also learned in this way: people will treat one as a member of the profession when one attends one's first professional meeting, presents one's first paper outside the department, submits one's first paper to a journal. Not all this latent learning is due to mentoring. Some of it is due simply to increased involvement in the social organization of the profession. But even the effect of increased involvement indirectly involves mentoring because the mentor is the link between education in a particular institution and the profession for which one is being educated.

The greater the integration of the mentor in the profession, the greater the integration achieved by socialization of a particular institution's students. Because so much of this kind of learning is latent, mentoring in this case is again largely modeling. But unlike modeling *how* to do something, this kind of modeling involves learning *who* to be. It is learning the characteristics of the model, learning an identity. A fact we will have to deal with later is that one of the unintended by-products of identity modeling is unfortunate: it involves learning irrelevant as well as relevant characteristics, including the race, ethnicity, and gender of the model.

Graduate education does not merely socialize students for a profession, it places them in it. Thus, universities are not only educational institutions, they play an important role in the labor market. One consequence is that mentoring involves not only personal and academic but also career advising. Mentors play a role in shaping aspirations and expectations for careers, they advise on such skills as how to write a vita, how to prepare papers for and submit them to conferences and journals.

Thus, mentoring is especially important to graduate education because graduate education markets labor, and sponsorship is a significant factor in labor markets; because graduate education is oriented to advanced, complex skills that create new knowledge, and tutoring and modeling are important to learning such skills; and because graduate education involves not only cognitive learning but also socialization to values, norms, practices, procedures, and attitudes of a profession, a significant factor in which is both modeling performance and modeling identity.

As implied, answering questions about mentoring is difficult because the term has so many meanings. It means a sponsor, a model, someone who is supportive, several kinds of advisor, and several kinds of supervisor. Depending on how one counts, there is anything from a half dozen to almost a dozen mentor roles. About all they have in common is that mentoring is almost always thought of as a one-to-one relationship, in the sense that one student can count on one particular mentor. How one answers almost any question about mentoring depends a good deal on how one answers two prior questions: must all these different roles be combined in one person for successful mentoring? Are they all necessarily one-to-one roles?

In my own view not all the roles of a mentor need be combined in one person. They are not even all one-to-one roles. I think these conclusions materially affect how one approaches questions such as how to encourage more mentoring, how to protect the independence of students, and how to match minority students to mentors. (What I have to say is therefore basically organized around these two conclusions. I precede them with a more elaborate consideration of the different roles involved in mentoring. I then explain why I do not believe all these roles are necessarily combined in one person, or are even all one-to-one roles. Finally, I address the implication of these two conclusions for the encouragement of mentoring, the independence of students, and the matching of minority students to mentors.)

WHAT MENTORING IMPLIES FOR THE STRUCTURE OF GRADUATE EDUCATION

In thinking about the implications of mentoring for the structure of graduate education it is first of all important to recognize that mentoring is a means to an end, not the end in itself. To think about its implications one has to take certain ends as given and then ask how mentoring relates to them. Research, for example, is an end, not merely the means, of graduate education. One reason that mentoring is important to graduate education is that it is a better means of learning how to do research. It is research that implies mentoring, not mentoring research.

Among the ends that I would like to take as given are, at the least, excellence, research, and autonomy. Given these ends, the more straightforward implications of mentoring for the structure of graduate education include a relatively early transition from course work to directed graduate teaching and research, a relatively small program, and highly selective criteria of admission.

Both teaching and doing research are best learned by teaching and doing research guided by a more experienced teacher or researcher. Both should begin as soon as possible. Not that all course work ceases. Seminars, for example, become more rather than less important. But they also should be organized around ongoing research. I have neglected the role that older students play in modeling behavior for newer students, but older students are models, too, and observing them perform in research-oriented seminars is an important part of a graduate student's growth. But mentoring not only implies early apprenticeship, it also implies a small population of graduate students. Because it is a one-to-one relation, it implies a high ratio of faculty to students, hence probably a number of students not much more than three or four times the number of faculty. This, in turn, implies selective admissions. While all universities desire the highest quality graduate students they can find, there are some that admit them in large numbers and sort them out only after the first or second year of work. The assumption is that undergraduate records and test scores poorly predict success as a researcher and, hence, as a graduate student. The assumption is sound, and the policy therefore rational, but it works against mentoring: mentoring implies more selective criteria because it implies a smaller number of entering students.

Because mentoring is such a complex combination of roles, there comes a point at which one cannot just take certain ends as given. Its implications are sometimes conflicting, depending on which role one has in mind. Different roles sometimes imply contradictory structures and their implications sometimes differ for different ends. To understand the implications of mentoring therefore sometimes requires thought about priorities among ends. Three such cases are the modeling of identity, the particularistic tendencies of sponsorship, and modeling autonomy.

The most controversial question about modeling identity is how similar a mentor has to be to his or her student in status characteristics such as race, ethnicity, and gender. Can a minority or female graduate student succeed if his or her mentor is a white, Anglo-Saxon, Protestant male? On the one hand, the answer is yes. Minority and female graduate students can learn how to do research as well by watching white, Anglo-Saxon, Protestant males as anyone else. It will of course immediately—and correctly—be objected that a female, say, is not going to learn from such a model how to combine family and career, and as long as role differentiation survives in the American family she needs to learn from a model certain things that only a female model can teach. But not all models are mentors, and what is needed here is someone to observe rather than tutorial direction. If, then, there is need for similarity of models to students—and I am about to argue that in fact there is such a need—it is for a different end, the implication of a different means. There is strong sociological evidence that the sheer number of models in a social environment affects the aspirations and achievements of minority and female graduate students because it affects their legitimacy. The sheer number of such models redefines aspirations and expectations, resulting in higher levels of achievement, and makes what at first looked like deviant behavior more acceptable. Thus, the assertiveness normal to male professors looks less like a chip on the female shoulder as more and more women become faculty. Thus, it is not that mentoring implies affirmative action. But affirmative action is nevertheless important because the number of minority and female faculty, over and above its role in mentoring, affects expectations for and actual levels of minority and female achievement by graduate students.

Another dilemma is that arising from the tendency of sponsorship to be particularistic. Particularism is a preference by a sponsor for his or her own students simply because they are his or her own students. It is a problem particularly for an academic marketplace that prides itself on universalistic standards of excellence. It poses a dilemma because more universalistic mechanisms of the market have their own problems. It is as difficult to predict successful academic careers from graduate performance as it is to predict success in graduate school from undergraduate performance. Such uncertainty is often dealt with by leaning heavily on the supposedly informed judgment of faculty whose excellence in research is itself well established. Because of its vulnerability to particularism, however, the trend for decades has been toward more impersonal, universalistic mechanisms of market allocation: public advertisement of positions, formal applications for them, bureaucratized interviews at professional meetings. Even more informal mechanisms have probably been broadened, that is, search committees consult more sponsors than they used to. The student benefits from wider diffusion of information and hence more options; employers benefit from larger pools of talent. None of this would imply a dilemma if it was not for the uncertainty of academic markets. But sponsorship is probably more a ritual than a technically efficient solution to market uncertainty. That is, I doubt that it in fact predicts talent very well. Therefore, if universalism is a goal, it has to be recognized that sponsorship detracts from rather than contributes to its attainment. We need to depend on it less and live with the resultant uncertainty.

Studies of the academic market (e.g., Caplow & Magee, 1965, Reskin, 1979; and more generally, see Granovetter, 1974) show that personal contacts are important sources of information about positions and directly affect placement in one's first job. Sponsors help in this process by introducing one to colleagues, writing recommendations, making phone calls. The success of their efforts depends a good deal on their access to networks, to invisible colleges of academics linked across schools by common orientations, methods, and theoretical approaches (Crane, 1972). Such networks are flows not only of ideas but also of information about jobs and they influence how jobs are filled. Both placement in one's first job and advancement from one position to another depends on having a mentor and on how well placed a mentor is in such networks.

But without a dissertation an academic's career is very limited. Hence, a mentor frequently is identified with the student's dissertation adviser. This adds a fourth kind of advisor to the meaning of a mentor and a second kind of sponsor. As an advisor, it involves guidance in selecting a topic, setting standards one ought to achieve and reasonable time limits to achieve them, and providing constructive feedback evaluating performance. But simply by taking a student on, the advisor performs a kind of sponsorship role. Sponsors affect not only placement in a market but also the allocation of scarce resources inside a department, such as research assistantships, teaching assistantships, and fellowships. In some cases even survival is at stake; a student managing to complete the degree only because there is someone on the faculty willing to assume responsibility for him or her.

In many instances the student doing research supervised by an advisor is also doing things *for* that advisor. That is, she or he is doing research that contributes to the advisor's own research. Where this research is funded, and is one of the ways a department supports its students, a new element is introduced into the mentor/student relation. Doing things for the mentor in itself introduces an exchange relation. Funding of the research transforms this exchange into a labor relation in which the mentor is an employer. This has its positive side in that it increases involvement on both sides of the relationship. The experience is more meaningful to the student because what she or he is doing really counts. It is more involving for the research supervisor because she or he has a real stake in the outcome. But at the same time it introduces a potential for exploitation which is among the principal down sides of mentoring. It is one of the things graduate deans worry about, especially issues such as credit for ideas, co-authorship in collaborations, and autonomy in formulating, executing, and using the results of the thesis. It is one reason why mentoring is not all wine and roses.

Most difficult of all is the dilemma of modeling versus autonomy. Autonomy is one of the goals of socialization by graduate school, but modeling implies mimesis and apprenticeship and sponsorship imply dependency. In the case of modeling the difficulty is more apparent than real, because the evidence shows that a student with two or more models often combines them in innovative ways. This is facilitated by constructing programs that cut across the shop organization that sometimes arises from concentrating students around their mentors, making it easier to multiply models (that is, make requirements general, make examinations independent of shop, make it easy to change shops, etc.). The dependency that goes with apprenticeship and sponsorship is more problematic. Long periods of dependency tend to infantalize students, decreasing their autonomy. One obvious solution is to reduce the duration of graduate training. There are still some schools where it takes 10 years to a Ph.D. A five-year program is a more likely way of promoting autonomy of the scholar. But even a five-year program is vulnerable to the effects of scarce resources on autonomy. Even faculty committed to encouraging the autonomy of their students may have difficulty giving them freedom of choice of dissertation topics if the faculty are using their own grant resources to fund dissertations. The increasing scarcity of funds, especially of dissertation grants, has been especially costly to graduate education, and nowhere more than in its effect on the autonomy of graduate student research. Given scarcity, we have had to make a choice between efficiency and autonomy, usually at the cost of the latter.

But the whole process of graduate education is stressful. It is not only the potential for exploitation, but also the fact that the outcomes of research are highly uncertain, professional socialization raises aspirations and expectations, and one-to-one feedback, even when constructive, is frequently critical. Research, because it is not a matter of learning what is already known, places students at the frontiers of knowledge. It necessarily involves high levels of uncertainty, risk, and therefore anxiety. The level of expectations is also much higher: one is a student, but one is expected to become a master. Furthermore, research supervision and dissertation advising involves a steady stream of possibly constructive but nevertheless often negative feedback; if all the mentor provides the student is positive feedback, the student is unlikely to learn very much. Because graduate schools are fraught with high levels of anxiety, graduate students need a high level of support, encouragement, and infusions of confidence if they are to survive. They have to learn that, however badly things are going, they are good and they will, eventually, learn. There is a difference between specific criticisms—"you did some particular thing wrong"—and diffuse support—"what you did was wrong, but you are good, you will learn." The most successful feedback is a combination of the two. Thus, support is an important role of the mentor.

In sum, depending on how you count, there is anything from a half dozen to almost a dozen mentor roles. At the least there are: advisor, research supervisor/tutor, model, sponsor, employer, and support. But there are four kinds of advisors—personal, academic, career, and dissertation advisors; two kinds of models—of performance and of identity; and two kinds of sponsors—internal and external. By this count there are upwards of 11 mentor roles. About all they have in common is that they all assume mentoring is a one-to-one relation. But is even this true?

HOW MANY MENTORS SHOULD ONE STUDENT HAVE?

Friedman's *Mentors and Supervisors* (1987) is perhaps one of the more comprehensive analyses of mentoring to date, but, for me, reading it raised two questions: Why assume that all mentor roles ought to be combined in one person? Why assume that any one student has one and only one mentor? Friedman's analysis appears to be the most informative single study of the subject. But it at the same time reveals a certain narrowness in the definition of a mentor that seems characteristic of much of the small literature on mentoring. Friedman does omit employment, but otherwise assumes that a faculty member is not a mentor if she or he does not combine support, all kinds of advising, sponsorship, research supervision, and modeling in one person. For example, students in one engineering department she studied were satisfied with their advising and research supervision but, because of the way research in the department was organized, did not think they had mentors. Nor did their faculty, nor Friedman herself. Research in the department was organized in teams, each with several faculty and several students. Because the relation was many-to-many, it was not a "mentor" relation (Friedman, 1987, p. 63). Other examples from other departments included cases in which faculty functioned as advisors and research supervisors but as not sponsors, or as advisors and sponsors but not as models. Hence, there were no mentors. It is not that Friedman was being arbitrary in this. It was how both the faculty and students she interviewed also thought. Furthermore, the same faculty, students, and Friedman herself assumed not only that a person was not a mentor if she or he did not combine advisor, support, research supervisor, sponsor, and model in one person but that any student must have one and only one mentor.

Some mentor roles do combine in one person. In questioning Friedman's first assumption I do not mean to overstate the case. I only mean to question that *all* of them combine in one person. The reason that some roles combine in one person is that the reputation of a sponsor depends in part on the dissertation of the student the sponsor recommends for employment. This is often, correctly, perceived to be the chief reward for mentoring. But its other side is that, for reasons found in the sociology of authority, it is very unsettling for a mentor to be evaluated for an outcome over which she or he has no control. (This is called the "compatibility" principle by Dornbusch & Scott, 1975. The same principle underlies such practical maxims as the unity of command.) For an authority structure to be stable, the compatibility principle implies that being a sponsor should go with being a dissertation supervisor, because the dissertation is the chief product of the student at the time she or he is first recommended for employment. For the same reason, the sponsor-dissertation advisor is likely also to be the student's career advisor.

The sponsor-dissertation advisor-career advisor is also likely to be the student's personal advisor, but in this case the driving force is not the theory of authority but simply the likelihood that the sponsor/student relation acquires an intensity that makes for perhaps a closer relation than the student is likely to have with most other faculty. The closer the relation becomes, the more diffuse it is likely to become, hence the more likely it is to be personal as well as academic or professional.

But not all mentor roles necessarily combine with the sponsor-dissertation advisor-career advisor-personal advisor roles. (Let's agree to refer to this simply as the "sponsor" role to save repeating such an inelegant compound.) Friedman found that not all sponsors were models, hence not mentors. But it is not clear that a sponsor who is not a model is therefore not at least one kind of mentor. And even if all sponsors were models, it would not follow that all models are sponsors. Research on vicarious conditioning in fact shows not only that multiple models are common, but they even have positive effects on innovation (Bandura, 1986). The same is true for research supervision. All dissertation advisors are research supervisors, but not all research supervisors are dissertation advisors. Research on career productivity suggests that, like multiple modeling, multiple research supervision has positive effects on long-term productivity (Reskin, 1979). (Reskin found that the quality of a department as a whole had a greater effect on long-term productivity than the quality of sponsors. She reasoned that this is because the quality of the faculty contributes positively to graduate student training.) Support too is something that is likely to be positively facilitated if other people besides the student's sponsor are supportive. For quite different, though less interesting, reasons many features of academic advising (e.g., monitoring progress) are often separated from other advising and vested in a chair of graduate studies.

Mentor roles are not all one-on-one any more than they all combine in one person. It might seem reasonable to suppose that, even if there are many different mentor roles, at least there is some common meaning that underlies them. And if there is some common meaning, it must be their one-on-oneness. But for essentially the same reasons that they do not all combine in one person, they are also not all one-to-one relations. Mentoring may mean individual, personal attention, but it does not necessarily mean that a student gets such attention from only one faculty member. If a sponsor may be a model, but not all models are sponsors; if a sponsor may be a research supervisor, but not all research supervisors are sponsors; if a sponsor may be supportive, but not all support is given by sponsors; then a student may have more than one model, more than one research supervisor, more than one support. Furthermore, in each of these cases a many-to-one relation appears to have positive effects (Bandura, 1986; Reskin, 1979.) Thus, the idea of recommending to graduate students that they may find that the "perfect" mentor is actually a team that they themselves put together makes good sense (Stanford Office of Graduate Studies, 1990, p. 7).

In sum, (1) there are many different mentor roles; (2) they do not necessarily combine in one person; (3) they are not even all one-to-one relations. Do these conclusions shed any light on the questions graduate deans frequently ask about mentoring?

TWO QUESTIONS ABOUT MENTORING

Two questions about mentoring appear to be perennial. First, mentoring is clearly important, but appears often to be the exception, not the rule and how does one encourage more of it? Second, too little guidance is like too little mentoring, a kind of neglect; but too much of it threatens the independence of the emerging scholar. How does one achieve the right balance of guidance and independence?

The Amount of Mentoring

Friedman (1987) found mentoring the exception, not the rule. No faculty and only 15 percent of the students she interviewed defined theirs as a mentor/student relation and many features of the structure of the departments she studied were unconducive to mentoring. But some of the issues Friedman raises are more "real" than others. Some are issues, such as number of contact hours, that clearly are real, but some are matters of how one defines the term "mentor."

The "real" issues are themselves of two kinds. First there are problems of literal neglect, that is, infrequent contact hours. Thus, as the number of students in a department increases, it becomes increasingly possible to match faculty to student interests but the decrease in the ratio of faculty to students decreases contact hours. Second, some structures are more conducive to one or another kind of mentor role than others. Thus, as the ratio of junior to senior faculty increases the likelihood of sponsor/student relations decreases because few younger faculty have the connections that make sponsorship work. It also matters how many of the graduate students in a department are full-time and professionally oriented. The chief reward for mentoring is the success of a student's career. Absent other kinds of rewards for mentoring, students not oriented to or likely to succeed in a career are unlikely to attract sponsorship.

For some of these "real" issues the nature of mentoring suggests, if not "solutions," at least what might be done to facilitate more mentoring. Thus, some mentor roles *are* one-to-one, which suggests that more mentoring is encouraged by increasing the ratio of faculty to students. This, in turn, is facilitated by more selective admissions and limits on graduate enrollments. On the other hand, there are limits to how far one can go in this direction. Friedman also found that, because mentoring cannot be forced, it is facilitated by the sheer billet size of a department. I speak, of course, as the former chair of a department that was too small, but there is a certain critical mass simply in the number of billets allocated to a department below which matching of mentors and students with mutual interests becomes very difficult. On the other hand, when it comes to some of the other ways in which structure affects mentoring, what one has to keep in mind is that there are many kinds of mentoring. Thus, the effect of age on sponsorship is not necessarily the same as it is on modeling, advising, or support. There are many mentor roles for which age is no barrier. There are also other kinds of mentoring of part-time and noncareer-oriented students. But in this case, the absence of rewards is a serious problem, though the problem has a solution. Universities need other rewards besides those that come from sponsorship, because sponsorship is not the only kind of mentoring it wants to encourage. Furthermore, these should not be confined to prizes, though prizes are important. Too often mentoring is not treated as teaching, and therefore does not count in measuring teaching loads. Even when individual directed reading, for example, is counted it is frequently unnoticed that mentoring is a particularly intensive kind of teaching. But often such teaching is not even counted. Advising and office hours, too, are treated as minimally meeting requirements rather than as positive activities to be rewarded. I have noticed in my own school that increasingly mentoring is taken into account in salary setting, promotion, and tenure, for example, by asking for letters from former advisees and doctoral students in the course of tenure decisions. This seems to me to be an important way of encouraging more mentoring.

But the "absence" of mentoring is as much a matter of how the term is defined as it is of "real" social conditions. Thus, both Friedman and her interviewees did not think of a faculty member as a mentor to a student if she or he was not academic advisor, personal advisor, dissertation advisor, model, and sponsor all at the same time and in each case not unless the faculty member was the only mentor of the student. The absence of mentoring was not always a matter of neglect, of few contact hours; it was sometimes a question of how the term mentor was defined. There were many cases of satisfactory advising; a fair number of cases of satisfactory sponsorship, if not modeling; of satisfactory dissertation, if not personal advising; of personal, if not dissertation advising; and research supervision by several rather than one faculty member. I do not wish to underestimate the real problems of mentoring in universities, but at the same time I believe a less narrow definition of the term will discover more mentoring than Friedman did and, even where the amount of mentoring remains a problem, will hold out more hope of a solution.

Is Mentoring a Threat to the Autonomy of the Student?

Too little guidance is typically neglect, not independence (Dornbusch, 1987). But too much is often thought to be a threat to the independence of the student. Modeling, sponsorship, employment, and supervision of research and/or the dissertation all pose threats of one kind or another to autonomy. Friedman (1987), for example, found that faculty and students alike feared the "cloning" of students by their mentors. Funding of research is often feared as constraining a student's choice of thesis topics (Stanford Office of Graduate Studies, 1990), and even without it students are often so dependent on their mentors in one way or another that mentors have a great deal of potential power over not only their careers but their ideas. Especially in the case of the dissertation, the mentor may, subtly, even unconsciously, manipulate the student's interests so that they match the research needs of the mentor because the student needs the advisor more than the advisor needs the student.

But understanding mentoring as many roles that are not all necessarily one-to-one substantially decreases the problem that balancing too little with too much mentoring might otherwise create. Modeling does not necessarily threaten cloning because it is frequently modeling of multiple mentors. Excessive concentration of all mentor roles in the hands of one person does concentrate power, hence dependence, and is a real threat. It is all very well that one advises mentors to allow autonomy, but the problem is as much structural as it is personal. A more useful corrective is to not concentrate all mentor roles in the hands of one person. Some roles cannot be divided without placing a student in a conflict relation, for example, sponsor and dissertation advisor, but there is room in any department for requirements that cut across shoplike research teams, and the involvement of other faculty with other experiences in training. In fact, by far the most difficult of the threats to autonomy to deal with is the problem of the funding of dissertations. Where the research a dissertation involves costs money to carry out, only the availability of external dissertation grants makes autonomy of the student feasible, and this kind of money seems to be decreasing rather than increasing in availability. Clearly we need more support specifically available for dissertation grants.

SUMMARY AND CONCLUSION

Mentors have not one but many roles. They are personal advisors, academic advisors, career advisors, dissertation advisors, sponsors, employers, research supervisors, they provide support and model performance and identity.

Some of these roles necessarily combine in one person, but most do not. Sponsorship, dissertation advising, and career advising combine in one person because sponsors will want some control over a product, the evaluation of which reflects on their own evaluation. Personal advising will often tend to combine with these three roles, simply because of the intensity of the relationship they create. But the remaining mentor roles are readily differentiated from these three and played by others. Thus, it is not necessary that each student have only one mentor; it is not even necessary that all the different types of mentor roles be one-to-one. In particular, having many models, research supervisors, and supporters in each case has positive effects on a student's graduate education.

Mentoring is especially important in graduate education because, more than in undergraduate education, graduate education is vocational training; its outcome is placement in a job; it is concerned more with creating than transmitting knowledge, requiring skills that are learned more by doing than by reading or listening to lectures; it is concerned as much with socialization to values, norms, practices, procedures, and attitudes as it is with cognitive learning, learned by participation in a specialized subculture linked to a larger social organization by a model.

Thinking of mentoring as one role played by one person, or as many roles that necessarily combine in one person, tends to obscure many issues in graduate mentoring. For one thing it tends to underestimate the actual amount of mentoring in graduate schools and to overestimate the structural barriers in the way of encouraging more of it. In particular, structural barriers to one kind of mentoring (e.g., sponsorship) do not necessarily stand in the way of other kinds of mentoring (e.g., support, personal advising, or academic advising). It tends also to overestimate the threat of too much mentoring to the autonomy of the graduate student. In fact, dividing mentor roles among several faculty tends to be one way to encourage the independence of student scholars. Finally, it tends to overestimate the importance of similarity in race, ethnicity, and gender in matching mentors to students. It matters considerably whether there are models on the faculty for female and minority students. They have positive effects on legitimacy, aspirations, and expectations. But it is not essential that models also be sponsors, supervisors, advisors, or even supporters. It is the sheer number of them on the faculty that matters.

REFERENCES

Bandura, A. (1986). *Social foundations of thought and action*. Englewood Cliffs, NJ: Prentice Hall.

Caplow, T., & McGee, R. (1965). *Academic market place*. Garden City, NY: Anchor.

Council of Graduate Schools. (1990). The doctor of philosophy degree: A policy statement. Prepared by the Council of Graduate Schools task force on the Doctor of Philosophy Degree, Elisabeth Traugott, Chair.

Crane, D. (1972). *Invisible colleges: Diffusion of knowledge in scientific communities*. Chicago: University of Chicago Press.

Dornbusch, S.M. (1987). The relation of parenting style to adolescent school performance. *Child Development, 58*, 1244-1257.

Dornbusch, S.M., & Scott, W.R. (1975). *Evaluation and the exercise of authority*. San Francisco: Jossey Bass.

Friedman, N. (1987). *Mentors and supervisors: A report prepared for the institute of international education*. New York: Columbia University.

Granovetter, M.S. (1974). *Getting a job*. Cambridge, MA: Harvard University Press.

Kuhn, T.S. (1962). *The structure of scientific revolutions*. Chicago: Chicago University Press.

Reskin, B. (1979). Academic sponsorship and scientists' careers. *Sociology of Education, 52*, 129-146.

Stanford Office of Graduate Studies. (1988). Graduate student academic advising: Guidelines for departments. Stanford, CA: Stanford University.

Stanford Office of Graduate Studies. (1990). Working with faculty: A reference for graduate advisees engaged in research and teaching." Stanford, CA: Stanford University.

Turner, R.F. (1972). *The doctoral dissertation sponsor: A study of role expectations*. Unpublished doctoral dissertation, Teachers College, Columbia University (cited in Friedman, 1987).

PART II

MENTORING APPROACHES AND MODELS

AN EXAMINATION OF EFFECTIVE MENTORING MODELS IN ACADEME

Olga M. Welch

For some time universities have viewed mentoring as critical to professional development and mobility. Within the higher education academy such mentoring most often occurs in the informal but special sponsorship that a graduate student receives from a senior professor. In this context the mentor becomes a role model, providing academic advice and eventually assistance in gaining access to the profession (Blackburn, Chapman, & Cameron, 1981, p. 315). In reporting the findings of several studies focusing on mentoring in higher education, Tillman (1995 and in this volume) suggests that the relationships could be psychologically supportive as well as professionally beneficial to the protégé (Hall & Sandler, 1983; McNeer, 1983; Wright & Wright, 1987; Sands, Parson, & Duane, 1991; Claque, 1992). Further, over two decades of research has shown that individuals can and do benefit from having supportive working relationships (Kram, 1985; Thomas, 1990). These relationships have frequently been referred to as mentor-protégé relationships, sponsor-protégé relationships, and perhaps most accurately, developmental relationships (Clawson, 1980; Kram, 1985).

Diversity in Higher Education, Volume 1, pages 41-62.
Copyright © 1997 by JAI Press Inc.
All rights of reproduction in any form reserved.
ISBN: 0-7623-0086-8

Despite these studies the literature contains no uniform agreement about (a) the roles of a mentor; (b) what differentiates a mentor from a sponsor; (c) how mentors select protégés; and (d) whether mentoring occurs differently and with different results for women and persons of color (Mertz, Welch, & Henderson, 1990). Thus, while research identifies the nature and dynamics of established mentoring relationships (how they occur in organizations at a variety of levels, as well as the roles a mentor might assume), there is less clarity about other important issues.

This lack of consensus about the definition of mentoring and the roles of the mentor has impeded the development of a clear conceptual framework within which to conduct further research. Nor is there sufficient evidence to confirm that the research and literature on mentoring are discussing the same phenomenon. This is particularly troublesome in examinations of mentoring between faculty and graduate students. The proliferation of these examinations (nearly 750 studies and increasing annually) illustrates the intense interest in mentoring relationships even as it highlights the lack of uniformity discussed earlier.

It is the elusive quality of mentoring which began the five-year, cross-disciplinary conversation about mentoring in higher education that this book represents. The conversation has ranged from descriptions of mentoring relationships in community college settings, to issues related to cross-race and cross-cultural mentoring, to mentoring for career advancement across organizational contexts, to the phenomena of "negative mentoring." In each case participants have sought to present their own research on mentoring in the context of an ongoing dialogue, revealing that while mentoring relationships share some commonalities, they also retain differences that underscore their complexity. This chapter presents an overview of mentoring in the academy. The first part of this chapter provides a review of the career and psychosocial roles of the mentor most often cited in the literature. The second part examines three studies of graduate mentoring in higher education (Holland, 1995, Heinrich, 1995; Schockett, Yosimura, Beyard-Tyler, & Haring, 1983). The third part offers the developmental model for career relationships (Henderson, 1990) which could be used to analyze mentoring relationships in academe. The chapter concludes with an examination of two studies with complications for developing a mentoring model in higher education aimed at increasing diversity.

ROLES OF THE MENTOR

Several researchers have identified the roles of mentors. Perhaps the best known of these are Kram (1980, 1985), Missirian (1980), and Phillips (1977), who have all described the roles a mentor might assume. These studies are considered seminal and form the basis of the work of Holland (1995), Heinrich (1995), and Schockett and colleagues (1983) discussed later.

Kram described a mentor in terms of two functions, career (where the mentor sponsored, coached the individual, as well as gave exposure, visibility, and protection) and psychosocial (where the mentor serves as a role model, counselor, and friend).

Tillman (1995) found these same roles replicated in her study of faculty to faculty mentoring relationships. In these relationships career functions prepare the protégé for career advancement and include *protection* (protection from assignment to committees that are particularly time consuming and which can detract from research and teaching agendas required for tenure); *direction* (providing direction in formulating research questions, the important questions to ask, as well as frank but confidential feedback, in addition to suggesting strategies for accomplishing specific academic tasks); and *sponsorship* (nominating the protégé for desirable projects, positions, and promotion, assuring that needed resources and career enhancing opportunities are available, and assigning projects that increase the protégé's exposure to and visibility with influential colleagues holding similar research interests).

Psychological functions enhance the protégé's sense of competence, identity, and work-role effectiveness. These functions include *role-modeling* (serving as an appropriate role model for attitudes, values, and behaviors); *acceptance and confirmation* (conveying unconditional positive regard for the protégé and his or her work); *counseling* (encouraging the protégé to talk openly about anxieties and concerns and giving the protégé support that facilitates socialization and helps in coping with the stress and work demands of the new faculty role). Thus, according to Tillman the faculty mentor can perform a variety of Kram's functions, ranging from socialization to emotional support (Tillman, 1995, pp. 3, 4).

Missirian (1980) identified similar roles for the mentor, suggesting that he or she could be a sponsor, coach, and even a peer depending on the context and the protégé's needs. Phillips (1977) reported the roles of mentors to be supportive bosses/partners, sponsors, invisible godparents, peer strategizers, role models, patrons, and favor-doers.

Despite the similarities in the roles identified by these researchers, the literature does not clearly define a single set of mentor roles. In fact, Fields (1988) analyzed 23 selected research studies and found that no one single role that was identified was accepted by all the researchers.

This ambiguity was also reflected in the research of Mertz, Welch, and Henderson (1990) in their study of career advancement mentoring in business, industry, and higher education. In describing their roles the mentors most frequently identified themselves as guides, exemplars, patrons, or sponsors. They also raised the following questions related to their decision to mentor:

a. Can I operate under the protégé's intense scrutiny of my behavior, ideas, style, and judgment. Intense scrutiny is uncomfortable. Am I willing/able to be explicit about everything I do and say?

b. If I teach the protégé all I know, will I lose some of my power?
c. Will I be judged by the behavior of the protégé? Will we be linked to one another in the minds of others?
d. What kinds of experiences should I provide to help the protégé grow, as well as to test his or her potential?

These questions demonstrate the complexity of mentor roles, as well as the idiosyncratic nature of mentoring relationships. Indeed, the researchers found that along with *competence* and *potential*, mentors used four additional factors in selecting a protégé. These were *organizational fit* (philosophical and physical), *risk* (the potential harm that might accrue to the mentor as a result of identification with the protégé), *predictability* (the feeling of assurance that the protégé's actions and reactions would fall within an acceptable range), and *payoff* (the potential gain/benefit that accrued to the mentor as a result of the relationship with the protégé). Across organizational contexts (business and industry, education and government), these factors defined for mentors whether potential proteges fell within or outside of the "comfort zone" needed for selection, a "comfort zone" which is illustrated in the study of graduate mentoring relationships (Holland, 1995), doctoral advising relationships between women (Heinrich, 1995), and the network mentoring model (Schockett, Yosimura, Beyard-Taylor, & Haring, 1983) described in the next section of the chapter.

MENTORING IN HIGHER EDUCATION

In his paper, "Paths to the Academy: The Faculty Development of African-American Doctoral Students" and his chapter in this volume (see Holland's chapter to read the complete work), Holland describes his examination of the characteristics of African-American doctoral student-faculty advisor relationships. Participants were students who had completed one year of doctoral study and individuals who had finished their doctoral work. Both groups were from two large research institutions in the midwest. The study included 43 participants, 23 of whom were still enrolled in doctoral programs and 19 who had completed their degrees. Of those currently enrolled in programs, four represented the hard sciences (physical science, engineering, and the life sciences), seven were from the humanities and social sciences, and 12 represented the field of education. Of the 19 students who had completed doctoral programs, three represented the hard sciences, eight were from the social sciences and humanities, and eight were from the field of education.

Using a grounded theory approach (Glaser & Strauss, 1967), Holland (1995) combined systematic data collection, coding, and analysis with theoretical sampling (Conrad, 1982, p. 241 as cited in Holland, 1995) in order to generate an integrated theory closely related to the data about categories of advisor/student mentoring relationships. He conducted in-depth interviews with each of the participants which lasted from 30 minutes to two and one-half hours and identified five types of relationships that African-American doctoral students have with their major advisors. These are summarized below:

1. *Formal academic advisement relationships* characterized by little interaction between the advisor and the student and a focus on guidance considered necessary for the student to advance through his or her program of study. Such relationships are nondevelopmental in nature since the advisor does not nurture or groom the doctoral student.
2. *Academic guidance relationships*, the most frequently experienced relationship by the study's participants (18), consisted of flexible and collegial contact between the student and advisor, supportive and understanding advisors who were concerned about the students' needs, and advisors who provided academically related guidance and advice to the doctoral student. There was little or no personal or social contact between the student and the advisor.
3. *Quasi-apprenticeship relationships* in which the advisor provided the student with academic opportunities not available to most doctoral students, invited the student to work on projects that advanced the advisor's research agenda, and guided the student with respect to the student's educational needs and completion of the research project. In this respect the relationship is developmental since the student receives the opportunity to be involved in an activity that is widely practiced in research universities.
4. *Academic mentoring relationships* are developmental relationships because the major advisor provides the student with individualized guidance and assistance aimed at preparing the student for academic life in higher education. These relationships are characterized by a broader range of opportunities for the student than those provided in the quasi-apprenticeship relationship. These developmental activities include those that are separate from the typical in-class assignments or projects such as teaching or research assistantships.

   Further, the advisor becomes a developmental role model, who discusses or displays for the student the political realities of departmental and university governance in addition to providing information about the expectations of faculty members at a large research university. Although not close friends, advisors in such relationships take a personal interest in the student and work closely with the student to enhance the student's academic preparation.
5. *Career mentoring relationships* were the most extensive that Holland identified. In these developmental relationships the advisor takes a direct

and purposeful role in preparing the student for faculty employment in higher education. Further, the advisor is active in providing the networks and socialization needed for the student to enter the profession. Finally, the major advisor takes a personal interest in the student and in his or her career success.

Implications of the Holland Study

In describing the implications of his study, Holland suggests that the African-American doctoral students he interviewed engaged in a variety of relationships with their major advisors and that these relationships played a significant role in the academic life, satisfaction, and career preparation of the students. Not surprisingly, Holland concluded that the quasi-apprenticeship, academic mentoring, and career mentoring relationships had the most significant impact on the participants in his study. This finding underscores the importance of these kinds of mentoring relationships to advancement in higher education. In his study of university administrators, Kouba (1984) found that 52 percent of college and university presidents he surveyed indicated that they had mentors while Fenske (1986) found that over half of the female chief academic officers of colleges of nursing and deans of colleges of education were mentored. In both studies participants reported mentoring experiences similar to the quasi-apprenticeship and career mentoring relationships described by Holland.

DOCTORAL ADVISEMENT BETWEEN WOMEN

A major element of the classic mentor-protégé relationship (Levinson et al., 1978) involves the high degree of identification between the parties. Individuals tend to identify with persons who are like themselves on salient identity group characteristics (Miller & Dreger, 1968). Since mentor-protégé relationships frequently require a high degree of sharing, the issue of identification becomes particularly significant. Kram (1985) found this issue of identification in her examinations of cross-gender mentoring in that female protégés often reported difficulty with seeing their male mentors as role models, an important psychosocial function of mentoring relationships (Thomas, 1990, p. 482).

In her phenomenological study of doctoral advisement, Kathleen Heinrich (1995) explored 22 women's relationships with their dissertation committee advisors. She was interested in how much these relationships resembled earlier relationships with parental figures. The study is important because gender and power continue to be inextricably linked in higher education (p. 448). Thus, women doctoral students are more likely to choose as dissertation committee members and chairs male professors with ranks and salaries, more power, influence, and professional connections rather than female professors.

Participants

Heinrich's study involved participants who were women between 35 and 45, European American, married with children, and who lived in suburban, middle-class neighborhoods. These women were employed full time during doctoral study and graduated in 1985-1986. Consistent with doctoral programs in education nationally, the "typical participant" in the study had a dissertation committee composed of male advisors who were between the ages of 45 and 55 and full professors. The women advisors were typically between the ages of 35 and 45 and at the assistant or associate professorial rank. At the time the study was conducted, few women advisors were available to choose from.

Findings

Two findings emerged from the participants with women advisors in the study: (1) a yearning for advisors to assume mentoring roles; and (2) deep disappointment when advisors were not mentoring. The central themes in the study related to power as advisees described the task and interpersonal dimensions of their advisement relationships. Two types of power were described by the advisees: *personal power* (which they termed the power accorded to any person by virtue of being human) and *legitimate power* (that power vested in the individual by the institution in the form of professional rank and status within the university).

"Whether advisors or advisees owned or disowned personal or legitimate power was related to how power was handled in advisory relationships which, in turn, determined the roles advisors and advisees assumed" (p. 450). The way power, task, and interpersonal relationships and interpersonal dimensions balanced as well as how conflict was resolved resulted in a typology (Table 1) of three reciprocal, advisor/advisee roles (1) *"power with"* relationships between professional friend advisors and colleague advisees, (2) *"power over"* relationships between iron maiden advisors and handmaiden advisees, and (3) *"power disowned"* relationships between negative mother advisors and good daughter advisees or between inadequate advisors and overadequate advisees.

Implications of the Heinrich Study

For Heinrich, the most salient implication of the study was collegial sharing of power with advisees which distinguished those advisors who were perceived to be engaged in mentoring. "The warm, professional friendships that these advisees described suggested that these teams dealt effectively with both power and conflict" (Heinrich, 1995, p. 450).

Table 1. Power Balance, Typology of Advisor-Advisee Roles, Balance of Task and Interpersonal Dimensions, and Style of Conflict Resolution of Advisory Relationship

Power Resolution	*Mentoring vs. Non-mentoring*	*Advisory Team Typology*	*Task Interpersonal Balance*	*Style of Conflict*
With	Mentoring	Professional Friend; Colleague	Balanced	Negotiation
Over	Nonmentoring	Iron Maiden Handmaiden	Task Oriented	Direct Confrontation
Disowned	Nonmentoring	Negative Mother Good Daughter	Interpersonal	Avoidance
		Inadequate Overadequate	Interpersonal	Avoidance

Source: Heinrich, 1995, p. 451.

Conversely, there were those relationships where women told stories of "silent betrayal" in which ineffectual women advisors stood by as advisees foundered and sometimes failed through lack of assistance with the byzantine bureaucracy of the university or were victimized by associate male and female advisors. These advisors betrayed advisees with their silence, while advisees simultaneously betrayed themselves by remaining silent about their needs as well as their feelings of frustration, disappointment, or anger with female advisors (p. 450).

Heinrich further found that 17 of the women in the study were able to identify similarities between parental and advisement relationships. "Participants' responses fell into two groups, those with (1) supportive relationships with female or male advisors who reminded them of supportive parenting figures, or (2) conflictual relationships with mothers" (p. 456). She concludes that if mentoring is envisioned as individuals involved in relationships with one another for the purpose of achieving a goal involving reciprocity, empowerment, and solidarity, then further study of mentoring relationships between women must be undertaken to learn more about how women advisors own and use their legitimate power, share power, and negotiate differences to nurture the professional growth of advisees. Additional studies of "silent betrayal" in mentoring relationships must also be conducted to provide more insight on this intriguing issue.

NETWORK MENTORING

Incorporating elements of Kram's psychosocial phases of mentoring, Schockett and colleague's (1983) network mentoring model focuses on the ways in which the mentor and protégé operate that are mutually affirming and empowering (Figure 1). In this model the *initiation phase* begins as the mentor provides education and role modeling for the protégé. The mentor's subsequent undertaking of the function of sponsoring involves the risk of greater commitment to the protégé which marks the onset of the *cultivation phase.* As mentor and protégé actively engage in the functions that emerge during the cultivation phase, their relationship grows stronger and correspondingly the overall width and breadth of the mentoring functions expand. As the mentor provides later-appearing functions, however, less time may be allotted to some of the earlier functions, which will eventually disappear altogether. The waxing and waning of functions continues during the *separation phase*, at which time ambivalence is experienced as mentor and protégé begin a process of psychological disengagement. By the time the relationship has progressed to the *redefinition phase*, the primary function of the mentor is one of moving from a transitional figure to a friend/peer.

This model, while incorporating Kram's psychosocial functions of initiation, cultivation, separation, and redefinition, reconceptualizes them to correspond in a manner which emphasizes the shifting nature of the mentor/protégé relationship. That is, at any time during the relationship the mentor may become a protégé, while the protégé becomes the mentor. Thus, "connected knowing" strategies are encouraged rather than those that support the more "traditional" roles of mentor and protégé discussed earlier. Generally, the ways in which mentors and protégés might join together to construct the experience have not received much attention. Although the literature, much of which focuses on mentoring relationships in business, industry, and higher education, has acknowledged the importance of consensus between the mentor and protégé on the goals and objectives of the relationship, an implied acceptance of the unequal distribution of power has existed in which the mentor identifies the "promising" protégé whom he or she then agrees to enculturate into the norms of the organization. This perspective ignores the protégé's "connections" which predate the mentoring relationship. It also assumes that altruism on the part of the mentor is the sole rationale guiding the selection of the protégé. The networks and resources of the protégé which the mentor might find attractive are overlooked.

More recent research on career advancement mentoring (Henderson, 1990; Mertz, Welch, & Henderson, 1990) suggests the importance of reciprocity in the mentoring relationship. In this research the authors found that potential mentors considered access to the networks of a prospective protégé a significant factor in their decision to mentor. Certainly this was true in examinations of more traditional same-sex mentoring relationships involving white males in which the perceived potential of the protégé to open previously inaccessible networks or to create new networks for the mentor constituted an important selection criterion.

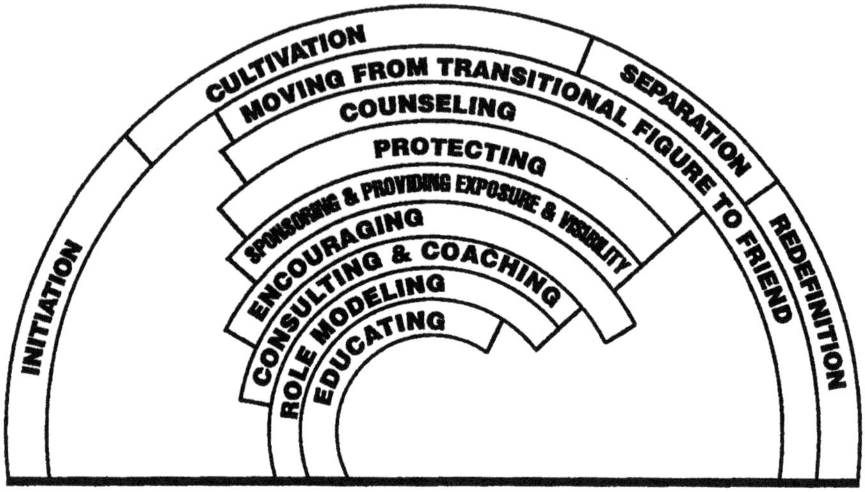

Figure 1. A Model of Mentoring

It is the protégé's co-equal ability to bring something of value to the mentoring relationship which is a major component of the network mentoring model. In a sense, network mentoring shares many of the characteristics described in the Holland and Heinrich studies. In both cases protégés who reported satisfying mentoring relationships were engaged in developmental associations which went beyond the concept of the mentor/protégé dyad. A developmental relationship is one that provides support for an individual's professional growth and career enhancement. It is also a relationship in which the parties have knowledge of one another and from which both may potentially benefit. Thus, developmental relationships are separated from connections to "heroes," after whom individuals may model themselves but with whom they do not have personal relationships, and from "temporary instrumental" relationships, which are of very short duration and require no mutuality between the parties (Thomas & Kram, 1987 as cited in Thomas, 1990).

The findings of the Holland (1995) and Heinrich (1995) studies of mentoring in the academy in conjunction with the network mentoring model (1983) offer important insights into the mentoring process. Another important addition to the literature on mentoring can be found in Henderson's (1990) study of career advancement mentoring in three organizations discussed in the third part of this chapter.

THE DEVELOPMENTAL MODEL FOR CAREER RELATIONSHIPS

Henderson's (1990) study of career advancement mentoring furnishes another conceptual framework within which to examine mentoring relationships in the academy. Although the study focused on career advancement mentoring, the findings regarding the perceived role(s) of the mentor in advancing the protégé to the highest levels of an organization can add to our understanding of mentoring in higher education settings.

Fifteen primary subjects participated in in-depth interviews intended to determine the perceived effect of a mentor on career advancement. For the study, a mentor was defined as "a senior level person in an organization who takes a personal interest in a junior member and facilitates that person's advancement by acting as a teacher, sponsor, role-model, or patron" (Henderson, 1990, p. 53). Across the three organizational contexts, (business and industry, higher education, and a governmental agency) the researcher's findings suggest that mentoring occurs with the most frequency in higher education. Further, the individuals at the highest levels in the business and industry and higher education institutions had multiple mentors. Moreover, only one-third of the subjects had intense, close relationships with the individuals they classified as mentors and none of those subjects held the highest positions in their organizations. Henderson used her findings to design the developmental model for career relationships (relationships) which differentiates four types of career assistance. These four types of assistance can be arrayed on a developmental continuum which distinguishes mentoring from non-mentoring and identifies the kinds of activities within each level.

As Henderson describes the model:

> Unintentional career assistance constitutes the most basic level (Level I-role model). Intentional passive career assistance constitutes Level II (a combination of role-modeling and the freedom to complete assignments and make some decisions). Neither of these is mentoring. Levels III & IV are mentoring. Intentional active career development mentoring occurs at Level III (including critical feedback, orchestrated experiences, and confidence building) And, intentional active career advancement mentoring occurs at Level IV (including the elements of Level III and sponsorship) (pp. 120-121).

Henderson concluded that contrary to prior research and writing, the individuals holding the highest position in each organization had not received career advancement mentoring (Level IV). Rather, they had received career assistance at Levels I, II, or III. The importance of career assistance confirmed in the study underscores the findings of the previous studies discussed earlier. Mentoring in higher education appears to be most beneficial when it is predicated on the mentor's commitment to a philosophy of developing people, not necessarily developing one particular person (Mertz, Welch, &

	NON-MENTORING		MENTORING	
LEVEL	I Unintential Career Assistance	II Intentional Passive Career Assistance	III Intentional Career Development Mentoring	IV Intentional Career Enhancement Mentoring
ACTIVITIES	role model	role model	role model	role model
		freedom to do assignment freedom to make decisions	freedom to do assignments freedom to make decisions	freedom to do assignments freedom to make decision
			career counseling confidence building critical feedback refining inter-personal and communication skills orchestrated experiences	career counseling confidence building critical feedback refining inter-personal and communication skills orchestrated experiences
				exposure to others sponsorship

Figure 2.

Henderson, 1988). This argues for additional research into the role of the mentor as sponsor and how the relationship facilitates career advancement.

Implications of the Models

The models presented emphasize the complex issues surrounding the term mentoring and its implementation in higher education. The work of Holland (1995), Heinrich (1995), Schockett and colleagues (1983), and Henderson (1990) offer interesting conceptual frameworks within which to examine mentor roles and protégé perceptions and expectations. Findings from each of the studies suggest that mentoring relationships in the academy can positively affect the graduate student (protégé's) predoctoral productivity and initial job placement (Scott, 1992 as cited in Tillman, 1995). These findings notwithstanding, there was still no uniform agreement about (a) the roles a person must assume for the relationship to be identified as mentoring; (b) the difference between a mentor and a sponsor; (c) how a mentor normally selects a protégé; and (d) whether mentoring occurs differently and with different results for women and minorities. Further, with the exception of the Schockett and colleagues (1983) model, the authors analyze grooming mentoring relationships in which the mentor assumes a variety of psychosocial and vocational roles intended to prepare the protégé for his or her position in the institutional context. For graduate students, grooming mentoring with its focus on an introduction to university culture suggests a "homogeneity of experience" (Lorde, 1984) among graduate students that is predicated on a "mythical norm" (Lorde, 1984, p. 115) which ignores the diversity of students' backgrounds.

Further, such one-to-one grooming models perpetuate a discourse on mentoring that concentrates on how to find the one mentor who can best facilitate the career of a protégé (Haring, 1996). Moreover, as the literature on grooming mentoring suggests (Levinson, 1970; Missirian, 1980; Mertz, Welch, & Henderson, 1988), such relationships place the mentor in a hierarchial position with respect to the protégé that can hinder mutual exchanges of benefits and support. Indeed, inherent in the grooming model is a disequilibrium in power that can impede full realization of the potential in the relationship for both the mentor and protégé. These and other issues connected with the grooming model of mentoring support the development of alternative models of mentoring suggested in the work of Henderson (1990) and Haring (1996). To build these alternative models, however, requires further consideration of cross-disciplinary definitions and examinations of the mentor/protégé relationship as well as the issues and questions that frame it. Such alternative models also can result in more effective mentoring of persons from underrepresented groups on university campuses as they seek to increase the diversity of their student populations.

ALTERNATIVE MENTORING MODELS TO ADVANCE DIVERSITY

In each of the studies those relationships which involved elements of cultivation (Schockett et al., 1983; Haring, 1996), quasi-apprenticeship, academic/career mentoring (Holland, 1995), or professional initiation in advising characterized by shared power between the advisor and advisee (Heinrich, 1995) were viewed by graduate students as positively affecting the quality of their doctoral experiences. In addition, Henderson's (1990) study emphasizes the variety and complexity of mentoring models for career development and the importance of a clearly implemented process to ensure their effectiveness irrespective of the organizational context.

Thus, according to these studies mentoring (despite its multiple definitions and complexities) offers some benefits during graduate study and in later career development. The work of Schockett and colleagues (1983), Haring, 1996), and Heinrich, (1995) particularly provide additional support for mentoring programs aimed at underrepresented groups on university campuses. Comprehensive mentoring models for underrepresented groups reflect the reality of being underrepresented and therefore less likely to find individual mentors (Haring, 1996). Although, for example, cross-race/cross-gender mentoring relationships can occur, the research reviewed here and elsewhere suggests that few African-American or women scholars experience success in these relationships (Frierson, 1990; Heinrich, 1995; Mertz, Welch, & Henderson, 1988). Given this reality, what factors could be considered in developing a comprehensive mentoring model that would be effective with underrepresented groups? The findings of the studies in this chapter suggest that such a model must consider and prepare students to operate within an organizational context (in this case a university) while recognizing the individuality of students' cultural backgrounds and perceptions of postsecondary education. In thinking about this model it is useful to examine the work of William Tinto whose study of student attrition offers some important insights into university context. Using Tinto's conceptual framework in his study of the college experience of Native Americans, Tierney (1993) sought to interpret the case history of one American Indian student within two contrasting theoretical frameworks (social integration and critical perspective). His use of critical theory to review Tinto's social integration theory can be useful when reflecting on the tendency of some mentoring programs to foster uniformity (e.g., cloning) without equal consideration to the conflicts such uniformity might engender in persons from diverse cultural backgrounds.

A second study with interesting implications for mentoring was conducted by Mickelson, Smith, and Oliver (1993) on the barriers in the academic hiring process facing African-American job candidates. Although the study examined the search process in higher education that resulted in fewer rather than greater numbers of African-American faculty at predominately white instructions, its findings regarding the twin barriers of sponsorship and meritocracy in the selection process can be useful in conceptualizing desired postgraduate goals of effective mentoring models for underrepresented populations.

THE UNIVERSITY CONTEXT

A comprehensive mentoring model must consider the realities of a postsecondary context. Tierney cites Tinto's (1984) model of student attrition, which he suggests rests on two ideas: (1) Van Gennep's rituals of passage and (2) Durkheim's theory of suicide, both of which assume a static reality. Van Gennep posited a series of rituals that facilitated the movement of individuals from one stage to another throughout their lives. Durkheim hypothesized that suicide occurred because of an individual's disconnection from larger societal structures such as the church, the state, and the family. This disconnection (or "anomie") becomes the catalyst for suicide.

Drawing on these two theories, then, Tinto suggests that student attrition represents not a failure of the student but rather the failure of the institution to integrate the student into the social and the academic life of the college or university. In his model, Tinto argues that some degree of social and intellectual integration must exist for the student to persist (Tinto, as cited in Tierney, 1993, p. 318). To accomplish this the student must commit a form of "cultural suicide" in which he or she replaces the values and assumptions of the home community (e.g., his or her "roots") in order to avoid intellectual suicide. As Tinto notes, "to become fully incorporated into the life of the college, they have to socially as well as physically disassociate themselves from the communities of the past. Their persistence in college depends upon their becoming departers from their former communities" (Tinto as cited in Tierney, 1993, p. 316).

This notation of "social suicide" as a prerequisite for social integration (defined by Tinto as a blend of the social and the intellectual) can have important implications for mentoring programs aimed at achieving diversity. It suggests that individuals from underrepresented groups who would successfully complete college must abandon (or choose to distance themselves from) their own cultural background and the perceptions of higher education within it that may differ markedly from those of mainstream society. As Tierney notes in his summation of Tinto's idea of social integration:

(In order for these individuals to achieve at the "ritual of college", they must accept the following the following assumptions about the postsecondary experience):

1. Postsecondary institutions are ritualized situations that symbolize movement from one stage of life to another.

2. The movement from one stage of life to another necessitates leaving a previous state and moving into another.
3. Success in postsecondary education demands that the individual become successfully integrated into the new society's mores.
4. A postsecondary institution serves to synthesize, reproduce, and integrate its members toward similar goals.
5. A postsecondary institution must develop effective and efficient policies to ensure that the initiates will become academically and socially integrated (Tierney, 1993, p. 316).

In critiquing these axioms, Tierney suggests that theory which offers explanations of current reality without also providing possibilities for the future fails to provide a framework for envisioning new paradigms. Thus, while acknowledging the significance of Tinto's findings regarding student attrition, Tierney (1993) calls for the development of new theoretical horizons that will permit researchers to consider multiple perspectives (voices) rather than using a unitary synthesis of data on what students should become (p. 317). He then offers a critical perspective on the five axioms above concluding that institutions that wish to increase their minority populations must create conditions which affirm and celebrate (and not with centers, or special courses, or periodic speakers) their lives. In such a setting minority students are encouraged to honor rather than abandon their histories even as they are urged to critically examine their own lives and the effects of society's forces on those lives. As Tierney states, "such a theoretical position has implications for virtually all of the organization—from how student affairs are organized to the manner in which knowledge is constructed, from the role of assessment to the role of the college president" (p. 322). In such an atmosphere inquiry on student attrition shifts from analyses which stress how those in authority can remedy the problems of those not in power, to investigations of power within the organization itself and how that organization can facilitate the development of empowering relations. In such an atmosphere individuals are encouraged to work in creative ways to transform the culture of the organization (Tierney, 1993). Moreover, within such a context the goals and outcomes of a postsecondary education that support integrative rather than inclusive practices can be examined in relation to the cultural dilemmas they present for students from underrepresented groups.

IMPLICATIONS OF THE TINTO AND TIERNEY STUDIES FOR DEVELOPING ALTERNATIVE MENTORING MODELS

As stated previously, the most effective mentoring relationships with graduate students are those that foster equalitarian rather than hierarchial exchanges between the participants (mentor/protégé). In studies of network mentoring conducted with graduate students from underrepresented populations in a variety of university settings, Haring (1996) found a resistance to the grooming model of mentoring (with its implied emphasis on assimilation and social integration). Instead, students and their mentors expressed a preference for network mentoring models because (a) they increased the availability of mentors, encouraging as they did reciprocity (e.g., mentors and protégés receiving "payoffs" from the relationship), (b) possessed equalitarian characteristics reminiscent of those found in the "power with" relationships described in Heinrich's (1995) study, and (c) allowed for flexibility and individuality as both mentors and protégés moved through the various stages.

These findings support Tierney's assertion that university programs which rely on data grounded in the implicit assumptions of integration rather than engaging in a critical analysis that makes central to discourse the students' own histories and backgrounds will inevitably subscribe to mainstream cultural assumptions about the desired requirements for successful college matriculation.

For graduate students, who must develop the kind of independent problem solving, critical analytical skills, and openness to a variety of theoretical and methodological approaches (both qualitative and quantitative) required for postgraduate work or careers, such a discourse is critical. Indeed, Holland's (1995) descriptions of the quasi-apprenticeship and career development mentoring relationships illustrate the importance of this kind of guided inquiry. In each of these relationships students and mentors engaged in the kind of critical dialogue and collaborative activities not possible in the traditional grooming mentoring associations.

Such a comprehensive mentoring model for members of underrepresented groups allows a greater potential for many psychosocial and vocational roles (Schockett & Haring-Hidore, 1985) to be played because of the involvement of more than one mentor. Haring (1996) suggests that grooming mentoring models often don't advance beyond the educating and role-modeling stages because of the inherent preferences and strengths of the mentors. With several mentors the possibility of completing the cycle more effectively is increased.

To be effective, then, a comprehensive mentoring model with underrepresented groups must foster the kind of equity (validation of both the mentors' and the protégés' experiences, histories, backgrounds, and critical perspectives) that encourages discourse, critical dialogue, and understanding of the role of power (within the relationship(s) and the institution). Thus, participants engage in an ongoing and evolving consideration of the desired goals and outcomes of the mentoring relationship, asking as Tierney does, "Are we seeing or hearing the multiple voices? And if so, what are the implications for action?"

MENTORING IMPLICATIONS FOR POSTGRADUATE WORK

A comprehensive mentoring model for underrepresented groups must consider the structural barriers existing in the labor market, regardless of potential or competence, that may impede students' acquisition of the postgraduate opportunities or employment in the academy. The study by Mickelson, Smith, and Oliver (1993) of barriers to the academic hiring of African Americans is illustrative. At the outset, the researchers argued that the barriers facing African Americans in academia bear some striking similarities to the obstacles they face in other labor markets. These similarities are especially critical given the current debate over affirmative action programs in the academy and a sample of the current data on black representation in higher education in the United States. As a recent issue of *The Journal of Blacks in Higher Education* (1996) reports:

- Not one of the 550 computer science faculty members of Harvard's departments of chemistry, mathematics, physics, biology, or computer science is black.
- Of the 160 faculty members in the political science departments at Dartmouth, Georgetown, Brown, Rice, Vanderbilt, or Berkeley, not one is black.
- Of the 919 professors on the physics faculties at the nation's 25 highest ranked universities, only seven are black.
- Of the approximately 75,000 U.S. medical school faculty in the United States, only 1,523 or 2 percent are black.
- Only 59 of the 978 members of English department faculties at the nation's 25 highest ranked universities are black.
- Blacks make up only 2.5 percent of all full professors at all American colleges and universities (p. 17).

Given these data, comprehensive mentoring programs must couple professional development objectives with attention to career development and the realities of selection and hiring barriers. Indeed Mickelson, Smith, and Oliver found that committees charged with hiring faculty utilize criteria in the search process that are elitist and exclusionary, resulting in fewer African-American candidates. The researchers found that the two indirect measures most often employed to determine the suitability of candidates (e.g., the social networks—referees—of the candidate and the prestige enjoyed by the institution from which the Ph.D. was conferred) constituted structural barriers for African Americans and disadvantages when seeking positions in the academy. Indeed, the findings suggest that far from supporting a meritocratic system, the measures resulted in the exclusion of minority scholars rather than their inclusion. As the authors note:

> In a quasi-objective manner, each discipline can claim to have knowledge of the best universities in a particular field. However, when institutional reputation is used as an exclusionary criterion, the end result is to unfairly exclude those Blacks and other minorities who do not have access to such prestigious institutions. (Further), because it is almost impossible for a student to gain access to a powerful and prestigious mentor unless he or she attends an elite university where these professors teach, African Americans are less likely to have such individuals as their referees (e.g., social networks).... In both instances, given the tendencies for these networks to be race, class and gender segregated, minority job seekers are disadvantaged (p. 15).

The study presents several implications for those interested in creating effective mentoring programs for underrepresented groups. First, it highlights the difficulty involved in assessing "merit" and the even greater difficulties posed for women and minorities in definitions which use the prestige of the institution as a proxy for merit in assessing their candidacies. This practice also underscores the problems involved in establishing criteria for such an assessment that are both meritocratic and "race neutral" (Mickelson, Smith, & Oliver, 1993). The researchers conclude with the call for a redesign of academic search processes that minimize, if not eliminate, the uses of institutional proxy as a measure of the candidate's merit.

IMPLICATIONS FOR A COMPREHENSIVE MENTORING MODEL

This research argues for attention to the relationship between career development (vocational roles fulfilled by the mentor) and the realities of the academic search process in designing mentoring programs. One strategy for accomplishing this, explored by Haring (1996), involves attention in the mentoring relationship to the analysis of organizational infrastructure. In each university context there exists a system that controls access to rewards. Explicating this context and involving protégés in the development of strategies for access is an important component of the cultivation and redefinition stages described in network mentoring.

Assisting graduate students to make the transition into the academy as colleagues involves a commitment on the part of mentors to have frank discussions of the application and interview processes including the structural barriers facing women and minorities. Simulations and problem-solving activities centered on issues of promotion and tenure can also assist in the graduate student's preparation for an academic career.

For women and minorities, this preparation is critical given the current affirmative action debates and the concomitant vigorous defense and endorsement of "merit" as the antidote to discrimination.

CONCLUSION

Educators in higher education considering mentoring programs should examine with a critical eye the information that we now have on the issue. At present it is clear that when we are talking about mentoring we are talking about any number of relationships occurring in divergent settings and with a range of purposes.

In order to evolve our understanding of mentoring and to conduct useful research on the subject, we must begin to clarify what we call mentoring and collaborate on what works and what does not. For those in higher education designing effective mentoring programs for underrepresented populations, the challenge could not be greater, or more important.

REFERENCES

Blackburn, R.T., Chapman, D.W., & Cameron, S.M. (1981). "Cloning" in academe: Mentorship and academic careers. *Research in Higher Education, 15* (4), 315-327.

Blackburn, R.T., Chapman, D.W., & Cameron, S.M. (1996, Spring). News & views. *The Journal of Blacks in Higher Education, 11*, 17.

Claque, M.W. (1992). *Promoting and retaining African American faculty: A case study of an aspiring multi-cultural research university.* Paper Presented at the Annual Meeting of the Association for the Study of Higher Education, Minneapolis, MN.

Clawson, J.G. (1980). Mentoring in managerial careers. In B. Derr (Ed.), *Work family and careers.* New York: Praeger.

Conrad, C.F. (1982). Grounded theory: An alternative approach to research in higher education. *The Review of Higher Education, 5* (4), 259-269.

Fenske, M.M. (1986). A comparison of the perceptions of mentoring relationships in the careers of female chief academic officers of nursing and male chief academic officers of education. Doctoral dissertation, George Peabody/Vanderbilt. *Dissertation Abstracts International, 47*, 07A.

Fields, W. (1988, Winter). Analysis of the concept mentor. *International Journal of Mentoring, 2*, 2, 12-19.

Frierson, H.T. (1990). The situation of black researchers: Continuation of a crisis. *Educational Researcher, 19* (2), 12-17.

Glaser, B., & Strauss, A. (1967). *The discovery of grounded theory.* Chicago: Aldine.

Hall, R.M., & Sandler, B.R. (1983). *Academic mentoring for women students and faculty: A new look at an old way to get ahead.* Washington, DC: Association of American Colleges Project on the Status and Education of Women.

Haring, M., & Welch, O.M. (1994). *Same-sex, cross-race mentoring for career advancement.* Paper presented at the American Educational Research Association, New Orleans, LA.

Haring, M.J. (1996). *Applications of the networking.* Mentoring Model in Formal and Informal Programs for Underrepresented Students. A presentation at the American Education Research Association Symposium: Minority Students, Mentoring Models and Mentors' Perceptions, New York.

Heinrich, K.T. (1995). Doctoral advisement relationships between women. *Journal of Higher Education, 66* (4), 447-469.

Henderson, J. (1990). Mentoring: Is it really necessary for advancement in business, education & government? Unpublished doctoral dissertation, University of Tennessee, Knoxville.

Holland, J. (1995). *Paths to the academy: The faculty development of African American doctoral students*. Paper presented at the Annual Meeting of the American Education Research Association, San Francisco, CA.

Kouba, N.L. (1984). The influence of helping relationships of the career development of college presidents. Doctoral dissertation, St. Louis University. *Dissertation Abstracts International, 46*, 07A.

Kram, K. (1980). Mentoring processes at work: Developmental relationships in managerial careers. Doctoral dissertation, Harvard University. *Dissertation Abstracts International, 41*, 05B.

Kram, K. (1985). *Mentoring at work*. Glenville, IL: Scotts & Foresman.

Levinson, D.J., Darrow, C.N., Klein, E.B., Levinson, M.A., & McKee, B. (1978). *Seasons of a man's life*. New York: Knopf.

Lorde, A. (1984). *Sister outsider*. Freeman, CA: The Crossing Press.

McNeer, E.J. (1983). Two opportunities for mentoring: A study of women's career development in higher education and administration. *Journal of the National Association of Women Deans Administrators and Counselors, 47* (1), 8-14.

Mertz, N., Welch, O., & Henderson, J. (1988). Mentoring for top management: Sex differences in the selection process. *International Journal of Mentoring, 2* (1), 34-39.

Mertz, N., Welch, O.M., & Henderson, J.M. (1990). *Executive mentoring: Myths, issues and strategies*. Newton, MA: Education Development Center.

Mickelson, R.A., Smith, P.S., & Oliver, M.L. (1993). Breaking through the Barriers: African American job candidates and the academic hiring process. In L. Weis & M. Fine (Eds.), *Beyond silenced voices: Class, race, and gender in United State's schools* (pp. 9-25). New York: SUNY Press.

Miller, K., & Dreger, R. (1968). Comparative psychological studies of negroes and whites. *Psychology Bulletin, 70*, 1-58.

Missirian, A.K. (1980). The process of mentoring in the career development of female managers. Doctoral dissertation. University of Massachusetts. *Dissertation Abstracts International, 41*, 08A.

Phillips, L. (1977). Mentors and protégés: A study of the career development of women managers and executives in business and industry. Doctoral dissertation, University of California, Los Angeles. *Dissertation Abstracts International, 38*, 11A.

Sands, R.G., Parson, L.A., & Duane, J. (1991). Faculty mentoring faculty in a public university. *Journal of Higher Education, 62* (2), 174-193.

Schockett, M.R., Yosimura, E.C., Beyard-Tyler, K., & Haring, M.J. (1983). *A proposed model of mentoring*. Paper presented at the annual meeting of the American Psychological Association, Anaheim, CA.

Schockett, M., & Haring-Hidore, M. (1985). Factor analytic support for psychosocial and vocational mentoring functions. *Psychological Reports, 57*, 627-631.

Thomas, D.A. (1990). The impact of race on managers' experiences of developmental relationships (mentoring and sponsorship): An intraorganizational study. *Journal of Organizational Behavior, 11* (6), 479-492.

Tierney, W.G. (1993). *The college experience of Native Americans*. In L. Weis & M. Fine (Eds.), *Beyond silent voices: Class, race, and gender in United State's schools* (pp. 309-324). New York: SUNY Press.

Tillman, L.C. (1995). *The mentoring of African American faculty: Scaling the promotion and tenure mountain*. Paper presented at the Annual Meeting of the American Educational Research Association, San Francisco, CA.

Wright, C.A., & Wright, S.D. (1987). The role of mentors in the career development of young professionals. *Family Relations-Journal of Applied Family and Child Studies, 36* (2), 204-208.

NETWORKING MENTORING AS A PREFERRED MODEL FOR GUIDING PROGRAMS FOR UNDERREPRESENTED STUDENTS

Marilyn J. Haring

The literature on underrepresented and at-risk youth has burgeoned in the last decade as well-intentioned administrators, faculty, and institutions have sought ways to recruit and retain these students. Programs have been instituted across the country that incorporate some or all of the following features to attract and keep underrepresented students: increased financial aid (e.g., Leon, 1993), specially designed courses to draw underrepresented students into the mainstream of academic life, leadership development, and of course, mentoring (e.g., Steele, 1991). "Of course" is used with mentoring because this concept and practice is frequently seen as a panacea in which underrepresented youth are molded to fit with college and transported to professional careers.

This chapter will examine two mentoring models that can contribute to the retention of underrepresented students. It will first examine the most common model employed in mentoring programs (grooming mentoring) and will cite

Diversity in Higher Education, Volume 1, pages 63-76.
Copyright © 1997 by JAI Press Inc.
All rights of reproduction in any form reserved.
ISBN: 0-7623-0086-8

examples where this model has guided mentoring programs in educational institutions. Then an alternative model (networking mentoring) will be discussed. Next, I will discuss a highly successful program which employed a strong component of networking mentoring for retention. Finally, based on my experience in designing more than a dozen mentoring programs, I will suggest key characteristics of successful networking mentoring programs and describe how these characteristics differ from those commonly utilized for grooming mentoring programs.

TWO MODELS OF MENTORING

In the literature on mentoring almost exclusive attention has been devoted to the grooming mentoring model. The chief characteristics of the grooming model are its emphasis on (a) one-to-one (i.e., dyadic) relationships, (b) hierarchy in which a mentor holds position and power over a protégé, and (c) unidirectional benefits which flow from mentor to protégé. In practice, based on this model, psychosocial and vocational assistance are offered by a person established within an institution or organization to a newcomer who is groomed and given special opportunities to succeed within the institutional culture (Schockett & Haring-Hidore, 1985). The net effect often is the maintenance of the institution or keeping the culture relatively intact, as expressed by Bullough (1996) in describing an archetypal grooming mentoring relationship: "Ultimately professors have stewardship over the soul of higher education" (p. 267). Figure 1 depicts the grooming mentoring model.

It should be noted that the psychosocial and vocational functions of mentoring are the basis of mentoring activities in both the grooming and networking models. Schockett and Haring-Hidore (1985) reported the results of a factor analysis in which the following factors clustered into what they called the psychosocial (or personal support) functions: (a) role modeling, (b) encouraging, (c) counseling, and (d) colleagueship. They also reported a vocational (career related) factor composed of the following functions: (a) educating, (b) coaching and consulting, (c) sponsoring and providing visibility and exposure, and (d) protecting.

Much discourse on grooming mentoring has focused on how to find the one mentor who can best facilitate the progress of a protégé by fulfilling one or more of the psychosocial and vocational functions. From the protégé's perspective, this discourse takes the form of, "How can I entice an important, well-placed person to take me under his or her wing?" From a programmatic perspective in academe, a focus is on where to find enough senior staff, faculty, or advanced students to pair with protégés who are underrepresented students. It also has taken the form of finding ideal ways to match mentors and protégés such that a chemistry will develop between the two parties based upon common

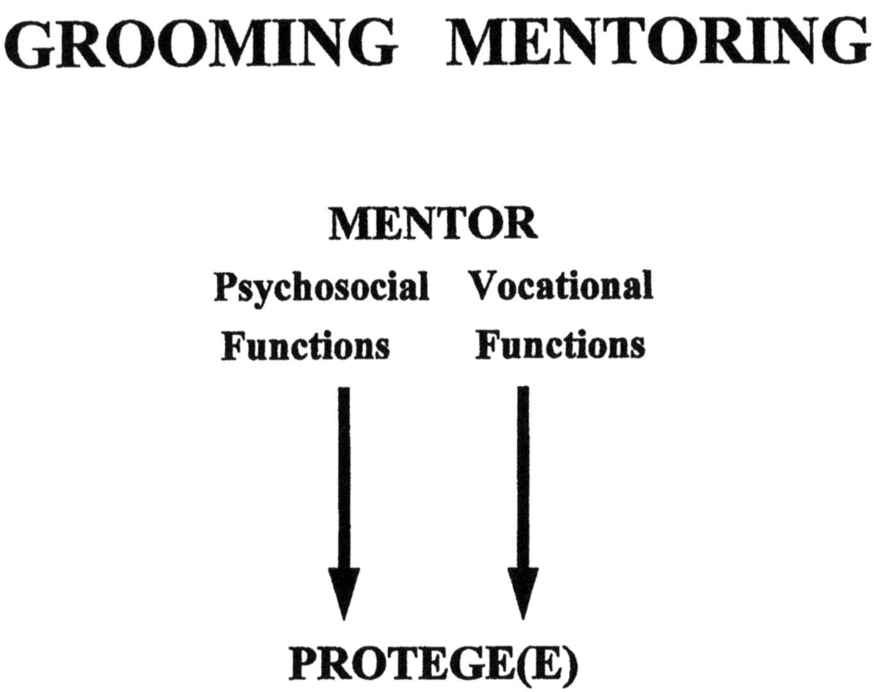

Figure 1. Model of Flow of Benefits in Grooming Mentoring

characteristics and interests. The bond formed by that chemistry is expected to facilitate a relationship which will be strong and enduring.

There are, however, three basic assumptions underlying the grooming mentoring model and the overriding concern for finding one magical mentor, which I vigorously challenge. One such assumption is that mentoring inherently is a dyadic relationship if it is to be of significant benefit to a protégé. Unfortunately, dependence on one person as an all-encompassing mentor has been shown to be perilous for some (Brooks & Haring-Hidore, 1987). In that study participants reported that mentors occasionally rejected their protégés; sometimes the mentors themselves suffered career setbacks which also negatively affected their protégés; and mentors often were perceived by protégés as deficient in offering their time, expertise, and encouragement. Thus, dyadic mentoring relationships often have flaws that can interfere with protégés' development and success.

A second assumption related to the search for the "right mentor" in the grooming mentoring model is that it is desirable to match a mentor and a protégé on some list of variables. These variables often include gender, race/ethnicity, age, academic major, and/or experience. However, there is no evidence to suggest that two people can be ideally matched for a mentoring relationship with any more success than two people can be matched for a blind date! In both cases there are intangibles which contribute to a chemistry that facilitates the relationship. People are not necessarily attracted to each other because they are both Hispanic, or are both from rural backgrounds, or both major in anthropology. Nor do people necessarily work well together because they share certain characteristics such as age or gender. These simple truths, however, usually are ignored when mentoring programs match people in a search for the right mentor for a protégé.

A third assumption underlying the grooming mentoring model and the search for a mentor is that there are clearly defined, fixed roles of *mentor* (senior person who is wise to the ways of achieving success) and *protégé* (inexperienced younger person who requires special support and education in order to be successful). This assumption constrains mentoring relationships inasmuch as the potential contributions which protégés bring to such relationships are ignored. As an example, a protégé may have valuable background, experience, or expertise in a variety of arenas which could benefit the mentor (e.g., living in another culture and/or being bilingual). But if the mentor is assumed to be in a superior position in all things, then the protégé is expected only to receive that which the mentor has to offer; and certainly a power imbalance that is inherent in the grooming model is exacerbated. In other words, equity has a small chance of occurring in such a relationship because programmatic focus is on what the mentor *offers* and what the protégé *needs*.

In recent literature on mentoring that focuses on retention of underrepresented students in postsecondary settings, a document prepared by the California Community Colleges (1993) is an especially rich compendium of suggestions for developing grooming mentoring programs for retention purposes. In addition to presenting 14 characteristics of mentors, these materials gave extensive mentoring tips to guide faculty to serve as mentors. The materials also described how to start a mentoring program for underrepresented students and included an annotated listing of 19 mentoring programs in California's community colleges. All but one of those programs appeared to be based on the grooming mentoring model. The lone possible exception was the peer mentoring program implemented by Santa Rosa Junior College (1993) in which underrepresented students were trained as mentors for newly enrolled underrepresented students. In that program, however, the dyadic relationships and benefits flowing from mentors to protégés were characteristics consistent with the grooming mentoring model.

Grooming mentoring, then, is the predominant model guiding mentoring programs today; however, I believe programs based on that model have not met the needs of large numbers of underrepresented students. The reasons for this distressing situation are numerous.

First, the very fact that they are underrepresented dictates that minority, disabled, women, and nontraditional students are unlikely to find the one mentor who will propel them toward their individual goals. There simply are not enough underrepresented mentors to match the demand. Also, members of the majority culture often gravitate toward protégés who are like themselves when they seek someone to groom, that is, these relationships tend to be what Dworkin (1996) termed "homogeneous."

In addition, there is a very real concern that mentors from the majority culture will unconsciously (or consciously!) mentor underrepresented students toward assimilation into the majority culture (Gonzales-Rodriguez, 1995). Underrepresented students whose needs are not met by assimilation either will not participate or not participate fully in such a mentoring program. The overall result of emphasizing only the grooming mentoring model, then, is that many underrepresented students never have—or utilize—the opportunity to participate in mentoring relationships.

Given the problems associated with programs based on the grooming model, my preference in working with mentoring programs is to encourage the use of an alternative model based on networking (Haring, 1993; Haring-Hidore, 1987; Swoboda & Millar, 1986). As will be discussed in the following paragraphs, networking mentoring is more inclusive. And it holds numerous other advantages for underrepresented people, who, because of their experiences, cultural perspectives, and values often are not comfortable with the concept and practice of grooming or cloning.

Figure 2 depicts aspects of the networking mentoring model. Among its characteristics are a number of people joining together to exchange psychosocial and vocational benefits, assisted by a skilled facilitator who

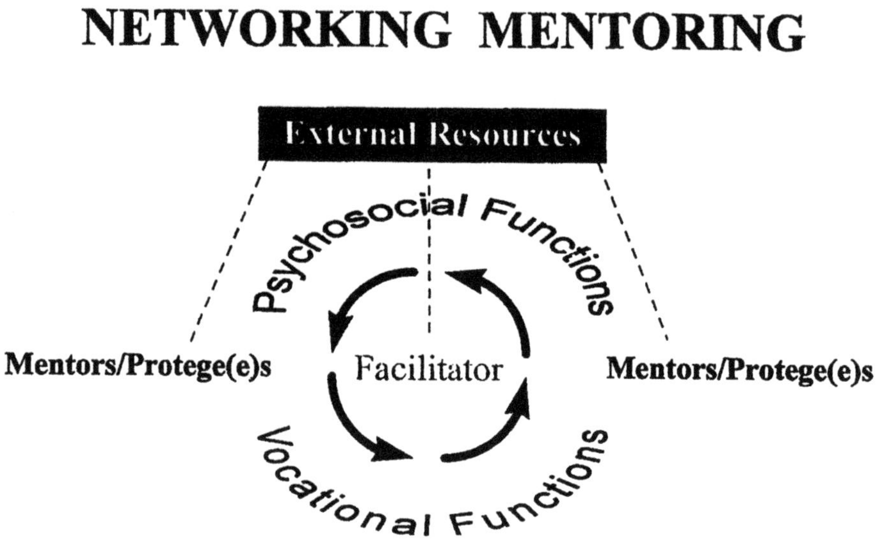

Figure 2. Model of Access and Flow of Benefits in Networking Mentoring

usually is assigned that role by a program or institution. And there is an intentional deemphasis of hierarchy and power within the mentoring relationship, such that at various times a participant in a network may act as mentor to others (when she or he has access to resources or has expertise to offer) and at other times may act as protégé (when desiring the encouragement, support, or assistance of others in the network).

Many of the advantages of the networking mentoring model are evident. For one, there need be no search for the perfect mentor who can offer all psychosocial and vocational assistance. Indeed, by networking with a group of three to eight peers, an underrepresented student has the opportunity to interact with a mix of people, no one of which is absolutely critical to support and success. In addition, the facilitator and the participants in the network have opportunities to access many different faculty, staff, and advanced students who can offer benefits that spread throughout the network (e.g., important educational information about the institution, projected career or major). At the same time, those senior people who are accessed do not have the opportunity to "clone" those in the network, and they are not so identified with the network that any possible downturns in their own careers will not have serious side effects for students in the network.

Another obvious advantage of networking mentoring is that no individual need be excluded—he or she has widespread opportunity to be a part of a networking group. And because there is little or no programmatic matching, there is reduced risk of poor chemistry negating the mentoring opportunity. Chemistry is less of a factor, also, because a group of peers who network can rely on less intense relationships than the very intense relationships which characterize the best grooming mentoring practice.

Further, because networking mentoring includes peers who are operating on a somewhat equal basis, there are reduced risks of assimilation pressures. In networking mentoring there are more opportunities to learn about various perspectives within the institutional culture, rather than a single perspective, as in grooming mentoring. In addition, it appears that with networking mentoring there is support for individuals to succeed within the institution but as part of a group which is likely to have more influence on the institutional culture. That is, grooming mentoring often contributes to a static institution because a protégé is groomed by a powerful mentor well established in the organization. Networking mentoring, on the other hand, has the potential to contribute to a more dynamic organization through the power of numerous people speaking with a single voice. If a network of underrepresented students questions the existing organizational culture, for example, they may well be taken seriously by those more senior because of the stronger voice achieved through greater numbers represented by the group.

An example of networking mentoring is a group of minority students who have been connected by the institution not only with each other but with alumni, administrators, and faculty within their university. Some of these connections probably were suggested and made by the group's facilitator who is experienced within the institution, focused on strengthening the network, and committed to supporting the development of those in the network. It should be noted that the networking mentoring facilitator is not a mentor but rather a support person to the network for organizing and connecting. The facilitator's focus is on the members of the network and facilitating their access to parts of the institution which can assist them as needed. He or she is free of obligations to perpetuate the institution's perspectives.

In this example, when the system in place does not foster their success in participating in campus governance, those in a networking mentoring group can propose a change in the way representation to councils and other bodies is determined. Their multiple connections and one voice (based on many) can pressure effectively for change of the system, which probably would not have resulted from the grooming mentoring model. In the latter model, the logical route to office in campus governance would have been through a solitary mentor providing visibility and exposure, encouragement, counseling, and education for an individual protégé who aspired to office. The underrepresented student may or may not have been able to achieve office, however, even with a powerful mentor if the institution were highly resistant to change. Thus, in this hypothetical situation the networking mentoring model would better serve individuals, the group, and I fully believe, the institution.

Although these advantages of the networking mentoring model are noteworthy, there are some critical factors which hold potential disadvantages that also should be mentioned. First, a networking mentoring program is more difficult to organize initially and to keep energized once it is started. Thus, a facilitator should provide needed support for organizing activities, meetings, and discussions, as well as seeking resources and accessing people within the institution that may be needed by the network. Further, a skilled facilitator is one whose organizing and connecting is more or less transparent. That is, she or he optimizes the participation of those for whom the network was developed and minimizes his or her importance. Thus, an unskilled or less responsible facilitator can effect some undesirable outcomes such as a weak network that is poorly organized or a network that is dominated by the facilitator.

Networking groups should be expected to have some ebb and flow, with members being more—and less—active across time, depending on demands upon them and their individual needs which can be met by the group. This makes it more difficult to monitor interactions and the meeting of individual needs in such a program because the dynamics are far more complex than in the dyadic grooming mentoring relationship. Thus, in order to achieve maximum benefits of the networking mentoring model, there is need not only for a skilled facilitator but also for each participant to make a serious commitment to give and receive benefits through the network.

In addition, it is critical that networking mentoring groups have access to senior people and other resources within the institution if they are to function effectively. Peer support is important and certainly is a fine benefit to those in a network, but the ultimate effectiveness of networking mentoring may rely upon the network actively accessing key resources within the institution and sharing with all participants that which is gathered (e.g., information, strategies, and backing). In this way, conversations are not simply internal among those underrepresented students engaging in the network; the dialogue is tempered by the reality of politics, process, and substance of the institution. It is the responsibility of the facilitator, in concert with the participants, to obtain access to people, knowledge, and facilities that are helpful to the network.

Finally, a potential disadvantage of networking mentoring is that it may not promote an individual as rapidly as grooming mentoring—especially for the lucky few who are selected by grooming mentors to be sponsored for astronomical promotion and rapid success. Nonetheless, a rich networking mentoring group can provide support for reaching career goals and can empower a group to impact an institution to better meet its members' needs.

A SUCCESSFUL PROGRAM UTILIZING NETWORKING MENTORING

Purdue University is a land grant state university which has made its strong reputation largely on the basis of its thriving engineering, agriculture, math/science, and technology programs. It is an institution where, in 1997, men still outnumber women (58% vs. 42%), despite strong efforts to recruit women into what has been for them nontraditional careers via the academic disciplines mentioned earlier.

As might be expected, the imbalance across genders in Purdue's School of Engineering is even greater than for the university as a whole. In 1996 male graduates with baccalaureate degrees in engineering outnumbered female graduates in those degree programs by a ratio of 3:1. At the master's degree level that year the ratio was five male engineering graduates to each female engineering graduate; and at the doctoral level the ratio was seven male engineering graduates to each female engineering graduate. Further illustrating aspects of this gender imbalance is the fact that among the 22 astronauts with Purdue engineering degrees, 21 are men and 1 is a woman.

In the past decade the Schools of Engineering at Purdue have made significant efforts to recruit and retain those students who are underrepresented in engineering, that is, women, minorities, and students with disabilities. Formal programs with mentoring components have been developed and implemented for these underrepresented groups; and one has been especially successful in employing an approach with many elements of networking mentoring: The M&M (Mentor and Mentee) Mentoring Program for Women (Wadsworth, 1995).

In the M&M Program first- and third-year women students in engineering were networked, as were second- and fourth-year women engineering students. One woman staff member facilitated each group of the 12 networked undergraduate women. Mentoring activities included informal meetings and discussions, as well as more formal seminars on topics of interest (e.g., graduate school). Often, women in this Sloan Foundation-funded program ate together and became better acquainted socially and professionally through these occasions. In keeping with the networking mentoring model there was no one-to-one matching of mentors and protégés in this program, as occurs in grooming mentoring. Also in the networking spirit, the principal investigator of this project ensured access of participants to Purdue officials (e.g., academic deans) as needs and interests arose.

The evaluation of the M&M Program revealed strengths and successes for a number of mentoring outcomes. As shown in Table 1, when compared to randomly selected control groups of women engineering students, participants in the M&M Program were more likely to be retained. Also, as shown in Table 2, participants agreed that the monthly meetings contributed to the goals of providing support, heightening self-esteem, and sharing strategies.

Table 1. Retention Rates for M&M Participants, and Controls

Group	*% Retention Rate*
Mentees	
1st Year	87
2nd Year	87
Mentors	
3rd Year	97
4th Year	100
Controls	
1st Year	79
2nd Year	72
3rd Year	78
4th Year	94

These outcomes appear to be related to the following mentoring functions described by Schockett and Haring-Hidore (1985): support (encouraging and supporting); self-esteem (counseling); and strategies (consulting and coaching). Further, according to the factor analysis reported in the study by Schockett and Haring-Hidore, two of these functions are psychosocial (encouraging and supporting; counseling) and one is vocational (consulting and coaching). One interpretation of the ratings given by M&M participants is that the mentoring program was rich both in perceived value and in type of mentoring functions provided.

KEY CHARACTERISTICS FOR SUCCESSFUL NETWORKING MENTORING PROGRAMS

Instructions for developing successful mentoring programs for underrepresented students commonly include tips that reflect assumptions of the grooming mentoring model rather than the networking mentoring model (e.g., Weinberger, 1992). Thus, instructions address careful matching of each protégé

Table 2. Participant Ratings of M&M Monthly Meetings' Contribution to Program Goals

	Mean Ratings		
M&M Group	*Support*	*Self-Esteem*	*Strategies*
1st and 3rd Year	4.0	4.4	4.6
2nd and 4th Year	4.3	4.0	4.5

Note: Ratings were on a five-point scale (1=strongly disagree, 5=strongly agree).

with each mentor, thoughtfully planned orientations so mentors can bestow benefits effectively, and prudent selection of professional people who act as mentors to underrepresented students, who also are wisely selected. Undoubtedly, this is programmatic grooming mentoring at its best.

Key characteristics for successful networking mentoring programs, however, differ from those just stated and, in fact, represent a different set of values, activities, and—I would argue—outcomes. The following are my suggestions for key characteristics of networking mentoring programs; no other list was found in a search of the mentoring literature.

Skilled Programmatic Facilitation

In successful networking mentoring programs there is a person assigned to the program (or who voluntarily begins a networking mentoring effort) who must be a skilled facilitator. Although there are organizers and directors in grooming mentoring programs, there is greater need in networking mentoring groups for a person who will facilitate an egalitarian group and will enhance its access to the institution. In addition, the emphasis in networking mentoring is not on directing the program so much as it is on gently organizing and guiding it. It seems to me that this touch is especially important when underrepresented students are the participants, that is, they may not respond well to insiders in the institution who direct them in inflexible ways. Instead, a network can be extremely supportive if it embodies the hopes, fears, styles, past experiences, present perceptions, and many strengths of underrepresented students. Such an effective network requires facilitation which is inclined and committed to nurture, with a minimum of emphasis on hierarchy and conforming to the status quo.

Provision for Access to the System

Another key characteristic of successful networking mentoring programs is that there is an infrastructure which gives underrepresented participants access to the system that controls the rewards they seek. Although this structure may vary, I suggest that there must be a commitment from well-placed persons in the organization to hear messages, transmit or act upon the messages, and support change for which the underrepresented group makes a strong case. Such well-placed members of the organization need not be underrepresented themselves, but often they are! Through their actions they give the network a voice and opportunities to impact the system. Perhaps more importantly, these advocates within the system give some measure of security, as well as encouragement, to members of the network who may not fit especially well within the present organization but who will invest in changing the system. Thus, in networking mentoring underrepresented students do not need to be clones in order to succeed, as is often expected in grooming mentoring.

Flexibility

In successful networking mentoring programs there is flexibility for redesigning activities, groupings, and degrees of support according to participants' needs. It is expected that the network will evolve and that new pursuits of the network will result from members' experiences and challenges at the institution. Related to this evolution is a need for flexibility in timing and frequency of meetings. For example, early in the program it may be desirable to have regular, planned gatherings; but later a more spontaneous schedule may emerge in which members of the network gather more (or less) frequently to educate, encourage, counsel, or seek and provide whatever mentoring functions are needed. It is desirable for members of the network to focus on their needs and then determine a schedule.

In addition, it is essential that networking participants continuously redefine their roles between mentor and protégé. When a person brings more knowledge or experience to an issue or activity than other members of the network, it is natural to assume a role that is more akin to mentor at that particular time. This is also true when a person assumes leadership for accessing resources and people on behalf of the network.

Real Caring That Leads to Individual Accountability

Participants in successful networking mentoring programs go beyond the formal boundaries established programmatically and give and receive benefits in caring ways. Thus, participants invest in each other and are conscious of the effect they have on each others' success. This leads to a sense of responsibility to the other participants and a willingness to be held accountable to contribute as well as to receive. This key characteristic, perhaps more than others that are descriptive of the networking mentoring model, must be addressed very intentionally by the facilitator of the mentoring group. This may require reminding group members of actions that are not accountable as those actions occur, for example, a member not sharing important information which would have benefitted others in the group.

Emphasis on Both Psychosocial and Vocational Mentoring Functions

By providing seminars or workshops that are mainly educational, many mentoring programs have focused on vocational mentoring functions. However, college students in Schockett's (1984) study expressed a statistically significant preference for the psychosocial functions, which provide personal support. Each time after that study when I have gathered data informally on this question, the same preference for psychosocial support has emerged. Thus, to fully meet the needs of participants in a networking mentoring group, I believe there should be programmatic focus on both vocational and psychosocial mentoring functions. This may be especially true for underrepresented students who are in institutions where they are newcomers. To consciously provide only (or even mainly) vocational assistance is to miss an important opportunity offered by networking, that is, linking in personally supportive ways those who often feel isolated in a new setting for members of their group.

CONCLUSION

For too long it has been assumed that "mentoring" and "grooming" are synonymous, with the alternative approach of networking mentoring receiving little or no attention. This omission of an important alternative model for programs to support underrepresented students is both noteworthy and regrettable. The assumptions underlying the two models are markedly different and contribute to contrasting expectations regarding participants' behavior, program characteristics, and outcomes.

Networking mentoring assumes the more egalitarian approach to supporting underrepresented students' success, even to the extent of promoting change in an institution's culture as a result of newcomers impacting it with their perspectives. In contrast, grooming mentoring almost always is the preferred model of established organizations and those charged with implementing mentoring programs for retaining underrepresented students. Its characteristics include hierarchy and a power imbalance (both skewed in favor of mentors). It has been utilized on behalf of achieving diversity goals for organizations, often with little regard for the complex needs of underrepresented students. I heartily recommend networking mentoring as a preferred model on which to base successful mentoring practice.

REFERENCES

Brooks, L., & Haring-Hidore, M.J. (1987). Mentoring in academe: A comparison of proteges' and mentors' perceived problems. *International Journal of Mentoring, 1* (2), 3-10.

Bullough, Jr., R.V. (1996). Professorial dreams and mentoring: A personal view. In C. Kridel, R.V. Bullough, Jr., & P. Shaker (Eds.), *Teachers and mentors* (pp. 257-267). New York: Garland Publishing, Inc.

California Community Colleges. (1993). *Student mentoring: Responding to the LaRoche challenge* (Report adopted by the Academic Senate). Sacramento, CA.

Dworkin, J. (1996). *The importance of mentoring in the promotion and tenure process.* Unpublished manuscript. Purdue University, West Lafayette, IN.

Gonzales-Rodriguez, Y.E. (1995). Mentoring to diversity: A multicultural approach. *New Directions for Adult and Continuing Education, 66*, 69-77.

Haring, M.J. (1993). Mentoring for research: Examining alternative models. In N. Minghetti, J. Cooper, H. Goldstein, L. Olswang, & S. Warren (Eds.), *Research mentorship and training in communication sciences and disorders, proceedings of a national conference* (pp. 117-125). Rockville, MD: American Speech-Language-Hearing Foundation.

Haring-Hidore, M.J. (1987). Mentoring as a career enhancement strategy for women. *Journal of Counseling and Development, 66*, 147-148.

Leon, D.J. (1993). *Mentoring minorities in higher education: Passing the torch.* Washington, DC: National Education Association (ERIC Document Reproduction Service No. ED 365 195).

Schockett, M.R. (1984). Relation among gender of mentor, gender of students, and mentoring functions. Doctoral dissertation, Arizona State University, 1984). *Dissertation Abstracts International, 45*, 1301-A.

Schockett, M.R., & Haring-Hidore, M.J. (1985). Factor analytic support for psychosocial and vocational mentoring functions. *Psychological Reports, 57*, 627-630.

Steele, R. (1991). *Mentoring: An effective tool for retention of minorities.* (ERIC Document Reproduction Service No. ED 342 841).

Swoboda, M.J., & Millar, S.B. (1986). Networking-mentoring: Career strategy of women in academic administration. *Journal of NAWDAC, 49*, 8-13.

Wadsworth, E.R. (1995). *Annual report to the Alfred P. Sloan Foundation.* West Lafayette, IN: Purdue University.

Weinberger, S.G. (1992). *How to start a student mentor program* (Phi Delta Kappa Fastback 333). Bloomington, IN: Phi Delta Kappa Foundation

MODELS IN MENTORING THROUGH FACULTY DEVELOPMENT

Marcia E. Canton and David P. James

This chapter discusses two distinct mentoring programs and models that have made positive impacts on the academic success of nontraditional students at an urban comprehensive community college, Prince George's Community College, in Prince George's County, Maryland, and at San Jose State University, a major California State University system campus. The focus is on students of color who historically demonstrated high attrition rates at those two schools.

THE PRINCE GEORGE'S COMMUNITY COLLEGE MODEL

In the past decades there has been increasing concern about declining enrollment of students in higher education (e.g., Brophy & Good, 1974; Tinto, 1987). Although the patterns of persistence vary by ethnic group, the major dropout point is the first year of college attendance. Coupled with this high rate of attrition during the freshman year, college's and universities must confront another compelling issue—the continuing decline in the pool of traditionally academically prepared freshmen (Bliss & Mueller, 1987).

Diversity in Higher Education, Volume 1, pages 77-92.
Copyright © 1997 by JAI Press Inc.
All rights of reproduction in any form reserved.
ISBN: 0-7623-0086-8

Increasingly, as the pool of traditionally prepared freshmen has declined, many institutions are admitting students who are academically underprepared for college level course work (e.g., Daloz, 1987). A Carnegie Foundation survey found that professors rate 75 percent of their students as lacking in basic skills (Rowe, 1981). A large proportion of these students are first-generation college students, many of whom are underrepresented minority students and/or from low-income families. For these students college may be overwhelming in terms of the intellectual demands and the persistent need to generate the needed financial resources to continue their education.

Often these high-risk students begin their freshman year with C-plus grade-point averages from their high schools and combined Scholastic Aptitude Test scores of less than 800. Some educators predict that the odds against this type of student successfully completing his or her undergraduate education are one in eight (e.g., Rowe, 1989). Universities and colleges have responded to the needs of these high-risk students through a variety of financial and academic programs which include work study opportunities, free tutoring services, and developmental programs.

Some research suggests that attention should be focused on assisting the high-risk student in becoming integrated into the campus social environment. Astin (1975) found in a national sample of college students, participation in these types of programs was associated with academic persistence. Astin emphasized the importance of students participating in the college or university nonacademic environment as a means of endangering the student's involvement and identification with the institution itself, and increasing the probability that students will persist. Tinto (1975) has reported that the decision to remain in school is a function of both academic and social success experiences. Success experiences can help the student to feel more closely integrated into the social and academic aspects of the university.

Some research suggests that mentoring programs can improve the academic performance and persistence of minority students (Clewell & Ficklen, 1986). Wide-ranging contact with faculty is an essential element to student life, an element which is critical in the formative years of the college career (Tinto, 1987). Other research suggests that freshmen benefit from mentoring programs which involve both faculty and upper-class students interacting with freshmen outside of the formal classroom (Tinto, 1987). Pascarella and Terenzini (1977) have found that informal student interaction with faculty outside the classroom may also be significantly related to persistence. Shapiro, Hazletline, and Rowe (1978) reported that non-persisters, when compared to persisters, had fewer personal relationships with students, faculty, or staff on campus.

BACKGROUND

Prince George's Community College in Prince George's County, Maryland, is a large comprehensive institution which serves the educational needs of a diverse, multicultural county whose population is 700,000. Minority enrollment at the college has dramatically multiplied over the past decade and a half from 9 percent in 1970 to the current level of 65 percent (1997), mirroring the county's surge in its nonwhite population.

Growing concerns about minority student performance prompted Prince George's Community College administrators and faculty to examine attrition patterns in the institution. In addition, with many minority students entering the college academically underprepared, the need for a comprehensive retention and mentoring effort became clear.

To combat the problem of minority student attrition and respond to the academic concerns of the underprepared, Prince George's Community College received funding in 1988 from the state of Maryland for the college's Black Student Retention Program. This grant proved to be a catalyst for a United States Department of Education Title III Grant, also awarded to Prince George's Community College in 1988, for a broad-based Minority Student Retention Program. Thus, a part of this chapter describes the retention and mentoring efforts and support undertaken by the college from the period of 1988-1991 to assist black and minority students with the skills necessary for success and retention.

OBJECTIVES

The general objectives of the black and minority student retention programs were to:

- increase the retention rate for first-time, full-time black male and female students.
- increase the retention rate for all other minority students (Asians, Hispanics/Latinos, Native Americans);
- increase faculty and staff sensitivity and skills in working with minority student populations;
- improve minority student adaptation to the campus environment; and
- develop a comprehensive minority student retention program that is fully integrated within the college's organizational structure.

SUPPORT SERVICES

Support is the linchpin of the black and minority student retention programs. The major supportive retention activities provided for both programs were:

1. mentors assigned to individual students;
2. academic monitoring of all program participants during the third, fifth, and eighth weeks of the semester;
3. career assessment and planning;
4. personal adjustment counseling;
5. workshops and learning fairs focusing on college survival hints;
6. tutorial services;
7. parent/student/mentor orientation programs each semester;
8. professional development activities to increase faculty and staff sensitivity and cross-cultural awareness; and
9. social and cultural activities promoting retention.

SELECTION CRITERIA FOR STUDENTS

Participation in the black and minority student retention programs was voluntary. To be eligible for support services provided by both programs, students were required to fill out a program application. In addition, all applicants were required to sign a Mentee Agreement Form which described the roles and responsibilities expected of student participants.

First-time, full-time black students were eligible for participation in the Black Student Retention Program. All other minority students (Asians, Hispanics, Native Americans), including part-time black students were eligible for participation in the Minority Student Retention Program.

MENTORING

The major element of service for students in the black and minority student retention programs was the individual support provided by mentors. All mentors were drawn from the college's full-time and part-time faculty, staff, and administrators. The mentors were selected by the project director based on the following criteria: (1) a successful record of working with students; (2) diverse backgrounds and interests; (3) professional training and experiences; (4) academic and personal counseling skills; (5) knowledge of the college and its resources; and (6) desire to serve as a mentor via the application process. The mentor's responsibilities include monitoring participants' academic progress, maintaining a supportive environment conducive to academic success, and initiating referrals to appropriate support service units such as the Writing Center, Tutoring Center, and Counseling Center.

Mentor-Student Matching

Mentor-student matching was primarily based on information listed on the student application such as: (1) major or program of interest; (2) skills

Table 1. Frequency Distribution of Faculty Mentors by Discipline, Gender, and Race/Ethnicity

Discipline	*(N = 10) Gender*		*Race/Ethnicity*		
	M	F	B	H	W
Afro-American Studies	X		X		
Biology	X				X
Communication	X				X
Education		X	X		
Engineering	X				X
Environmental Studies	X				X
English	X	X		X	X
Mexican-American Studies	X			X	

development; (3) interest in student application such as tutoring, study skills counseling, career assessment/planning, and (4) assistance in achievement of personal and career goals.

Mentoring Workshops

All mentors participated in mentor training workshops each semester. The workshops focused on effective mentoring techniques designed to assist mentors in developing positive relationships with minority students. Additionally, the workshop training focused on how to develop supportive classroom environments for minority students and prescribing for them, appropriate support service referrals within the college.

Instructional Support Services

Students selected for program participation also receive direct instructional support from trained staff members of the College's Tutoring Center,

Table 2. Frequency Distribution of Faculty Mentors by Gender and Race/Ethnicity*

	WM	*WF*	*(N = 10)* *BM*	*BF*	*HM*	*HF*
N	4	2	0	2	0	2
%	40	20	0	20	0	20

Note: *WM = white male; WF = white female; BM = black male; BF = black female; HM = hispanic male; HF = hispanic female.

Writing Center, Counseling (Career Assessment Center and Vocational Support Services). Instructional support personnel, including tutors, who were selected to participate in the retention programs also participated in training activities similar to that provided the mentors.

RESULTS AND OUTCOMES RELATED TO STUDENT PARTICIPATION

Black Student Retention Program

From its first semester of operation (spring 1988) to the spring 1991 semester the program grew from 34 students to an enrollment high of 346 students in fall 1989.

Spring 1988	34
Fall 1988	193
Spring 1989	216
Fall 1989	346
Spring 1990	261
Fall 1990	303
Spring 1991	280
TOTAL:	1,633

Minority Student Retention

From its first semester of operation (spring 1989) to the spring 1991 semester the program grew from 92 students to a high of 298 students in fall 1990.

Spring 1989	92
Fall 1989	221
Spring 1990	183
Fall 1990	298
Spring 1991	290
TOTAL:	1,084

Mentoring

The number of mentors providing individual support to program participants in both retention programs increased from 11 in spring 1988 to a high of 103 in fall 1990.

RETENTION RATES

Black Student Retention

From spring 1988 to fall 1988 the retention of Black Student Retention Program participants was 71 percent—12 percentage points above the average retention rate for black students. Participant retention from fall 1988 to spring 1989 rose to 80 percent. From fall 1989 to spring 1990 participant retention rose to 83 percent, which exceeded the average rate for any identifiable student population subset on the campus.

Spring 1988-Fall 1988	71%
Fall 1988-Spring 1989	80%
Spring 1989-Fall 1989	69%
Fall 1989-Spring 1990	83%
Spring 1990-Fall 1990	73%
Fall 1990-Spring 1991	80%
OVERALL RETENTION RATE:	76%

Minority Student Retention

The spring 1989 to fall 1989 retention of minority student program participants was 69 percent, five percentage points above the average retention rate for minority students. The overall minority participant retention rate from spring 1989 to spring 1991 was 68 percent.

Spring 1989-Fall 1989	69%
Fall 1989-Spring 1990	65%
Spring 1990-Fall 1990	66%
Fall 1990-Spring 1991	73%
OVERALL RETENTION RATE:	68%

ORIENTATION PROGRAMS

Orientation programs at PGCC were developed to provide structure to the relationship of mentors to mentees and parents of participating students. The orientation program, which was tested initially during August 1988, focused on the nature of the mentoring relationship, the definition of academic and career goals, collegiate classroom expectations, orientation to campus support services resources, and individual student problem-solving skills. Participants at the August 1988, 1989, and 1990 orientations numbered more than 300, including student participants, parents, and faculty/staff mentors.

Table 3. Frequency Distribution of Black and Hispanic Mentor Program Student Applicants by Class Level

			(N = 311)			
Class:	*Frosh*	*Soph.*	*Jr.*	*Sr.*	*Grad.*	*Undecl.*
N	69	45	116	66	7	8
%	22.1	14.4	37.2	21.2	2.2	2.5

IMPACT OF THE RETENTION PROGRAMS

The application of the black and minority retention program models proved to be successful for the participating students and Prince George's Community College overall. The following outcomes were observed:

1. increased graduation and transfer rates;
2. increased retention rates;
3. decreased attrition rates;
4. higher course completion rates;
5. increased grade-point averages;
6. expanded faculty/staff development programs for mentors;
7. improved orientation programs for students participants and parents;
8. positive student and mentor evaluations of the retention programs;
9. integration of the black and minority student retention programs in the college's general fund budget; and
10. positive nationwide publicity of the college's retention and mentoring programs.

CONCLUSION

In the October 16, 1989 issue of the *U.S. News and World Report*, Prince George's Community College was Maryland's only community college to be included in "America's Best Colleges" by junior and community college experts from across the country. The recognition cited the college's accomplishments in black and minority student retention.

In October 1990 the Black Student Retention Program, with its mentoring component, received the Distinguished Award of Merit in the Student Services Category from the Maryland Association of Higher Education. Additionally, in 1990 the Prince George's Community College Black and Minority Student Mentoring/Retention Programs received the Retention Excellence Awards from the Noel/Levitz Centers for Institutional Effectiveness.

In fall 1991 the college renamed the program ALANA designed to serve (African, Latin, Asian, and Native American students). The program today offers services such as college orientation sessions, faculty and staff mentors, visits to prospective institutions from where students can transfer, monthly newsletters with motivational and informational articles, cultural field trips, social receptions, and referrals to other support areas for additional assistance.

In developing a model for a comprehensive planned mentoring program, Canton and James (1994) lists 14 essential components to be considered for success in mentoring and retaining nontraditional students. These components, listed below, formed the foundation for ALANA.

1. A centrally planned and managed mentoring program office charged with responsibility for operating and/or coordinating the institution's mentoring activities.
2. Sufficient support staff compatible with the goals and objectives of the mentoring program.
3. A sufficient and clearly planned budget compatible with the goals and objectives of the mentoring program.
4. Clearly defined responsibilities of the administrative staff.
5. Clearly defined objectives of the mentoring program.
6. Clearly defined roles of the mentors.
7. Criteria for selection of mentors.
8. Criteria for selection of proteges.
9. Clearly defined responsibilities of the protégés.
10. Carefully designed orientation programs for protégés and mentors.
11. Effective matching of mentors and protégés.
12. Periodic discussion and review of literature with the mentoring staff focusing on effective mentoring techniques.
13. Marketing the mentoring program.
14. Program evaluation.

THE SAN JOSE STATE UNIVERSITY FACULTY MENTOR PROGRAM

San Jose State University's (SJSU) Faculty Mentor Program was developed to provide a supportive academic environment to reduce the escalating attrition rates among underrepresented minority and first-generation college students. Campus student retention data showed the highest attrition rates to be among African-American and Raza/Latino freshmen. Thus, these two groups became the initial target population.

In the fall of 1987 the campus was awarded lottery funding with matching funds from the campus to begin a concerted effort to ameliorate this problem.

A selection committee was formed to review the faculty applications. This committee was composed of the Faculty Mentor Coordinator, the Associate Dean of Academic Advising and Retention, the Student Affirmative Action Retention Coordinator, and the Associate Director of Equal Opportunities Program. Faculty selection was based on the following criteria:

1. demonstrated commitment to underrepresented minority students;
2. status and respect from colleagues; and
3. academic discipline to reflect anticipated disciplines of student participants.

As a result of this process 10 senior faculty mentors were selected for high-ratio mentoring (10 students to one senior faculty). To accommodate the large number of student applicants, the Academic Vice President also sent a general information invitation to faculty encouraging them to volunteer. One hundred San Jose State faculty members responded to this invitation and they were matched with 100 students. Two hundred students were therefore served by this program: 100 students with the senior faculty mentors (1:10), and 100 students with the volunteer faculty mentors (1:1). The remaining 300 applicants were notified that they were being placed on a waiting list until they could be matched with a volunteer faculty mentor. They were also invited to participate in all Student Affirmative Action (SAA) EOP (Equal Opportunity Program) activities. Their names were placed on the SAA mailing list and they were apprised of all events.

PHILOSOPHY AND GOALS

The faculty mentor program's philosophy and goals were consistent with the mission and purpose of San Jose State University. The philosophy, mission, and goal statements commonly recognized San Jose State University as a multipurpose institution of higher education located in a large and culturally diverse urban setting. As a result, the faculty mentor program was committed to providing resources to all students. It did not confine itself to any single approach to the accomplishment of this mission. It raised campus awareness about the importance of mentoring for a culturally diverse student population; it established training programs to help faculty learn or, in some cases, relearn how to be effective mentors for students; and it offered seminars and other programs to help students explore academic and career options indicated below were developed to address the needs of the program on three levels: (1) program goals, (2) faculty goals, and (3) student goals.

Program Goals

1. To establish personal contacts between faculty and students so that each may become aware of learning opportunities and problems.
2. To create a better understanding of culturally diverse populations for the students, faculty, and university at large.
3. To integrate the mentor program into the entire retention plan of the university.
4. To develop a training program for new faculty.
5. To evaluate the effectiveness of the mentor program for both students and faculty.

Faculty Goals

1. To provide assistance in solving the student's immediate problems through mentor counseling.
2. To inform students of available learning resources in the university.
3. To interpret the university culture to the students.
4. To collect data on resources to inform students, and to determine needs for the development of new resources.
5. To be active advisors and advocates for students.
6. To increase awareness of faculty about cultural diversity and student needs, and to communicate this awareness to the faculty at large.

Student Goals

1. To empower the student protégés to increase their effectiveness as students through knowledge of the culture and resources of the university.
2. To assist the students to evaluate their learning skills and to begin a program of improvement.
3. To assist the students in personal growth about cultural awareness, problem-solving techniques, and decision-making processes.
4. To help students establish long-term goals and objectives and plans for their accomplishments.

FACULTY MENTOR ACTIVITIES

Research on retention reports the importance of faculty interacting with students in a nonacademic environment. Toward that end, several social events were planned for the fall semester. These events were structured around the ideas and efforts of the faculty mentors.

Table 4. Distribution of Black and Hispanic Faculty Mentor Program Student Applicants by Class Level

	(N = 311)		
	Black	*Hispanic*	*Other*
Freshmen			
N	32	37	0
%	10.2	11.8	0
Sophomores			
N	21	24	
%	6.7	7.7	0
Juniors			
N	33	82	1
%	10.6	26.3	0.3
Seniors			
N	25	41	0
%	8.0	13.1	0
Graduate Students			
N	3	4	0
%	0.9	1.2	0
Undeclared			
N	2	1	5
%	0.6	0.3	1.6

Faculty mentor training was consistent with the overall goals of the program as reflected in the following mission statement:

> To create a better understanding of culturally diverse populations for the students, faculty, and the university at large.

Three training workshops were scheduled over a two-month period. Renown experts on mentoring assisted the participating mentors and protégés in developing goals, objectives and specific activities, understanding the cultural dynamics of African-American and Raza/Latino students. Outside community groups and students from local colleges attended the workshops on cultural issues. To foster a better relationship between schools and coordinate mutual efforts, this type of participation should be encouraged and continued in the future.

Mentors and students were expected to meet at least once a month. The Faculty Mentor Coordinator monitored their progress through telephone calls and questionnaires. At the end of the semester the faculty mentors met to provide a summative evaluation. Their recommendations were to:

1. Continue to provide three units of release time to senior faculty mentors. This serves as an incentive to participate and it also affirms the university's commitment to the program.
2. Clearly integrate faculty mentor activities with the retention, tenure and promotion process (i.e., faculty mentoring should be considered as a viable and professional contribution to the university).
3. Both senior and volunteer faculty applicants should be formally interviewed prior to their acceptance into the program. Although the faculty's intentions may be worthy, their attitudes toward black and Hispanic students should be assessed.
4. Interview all students who wish to participate in the program. This time should be used to acquaint them with the program and to advise them of their role in the mentor-protégé relationship.
5. Place the student and faculty data on a computer to facilitate accessing and monitoring information.
6. Develop a course with academic credit for protégés who participate in the faculty mentor program. Students should be rewarded with academic credit for their involvement with the program. This would provide an additional incentive for students to participate.

LESSONS LEARNED

The *purpose* and *goals* of a mentoring program should be clear, comprehensive (addressing goals for the program, mentors, and protégés), and familiar to all who participate in the development of goals is an effective strategy for soliciting support of participants and creating a sens the semester the faculty mentors met to provide a summative evaluation. Their recommendations were to:

Table 5. Gender Frequency Distribution of Selected Students by Race/Ethnicity

	(N = 99)		
		Race/Ethnicity	
	Black	*Hispanic*	*Other*
Male			
N	30	11	1
%	30.3	11.1	1.0
Female			
N	31	26	0
%	31.3	26.2	0

1. Continue to provide three units of release time to senior faculty mentors. This serves as an incentive to participate and it also affirms the university's commitment to the program.
2. Clearly integrate faculty mentor activities with the retention, tenure and promotion process (i.e., faculty mentoring should be considered as a viable and professional contribution to the university).
3. Both senior and volunteer faculty applicants should be formally interviewed prior to their acceptance into the program. Although the faculty's intentions may be worthy, their attitudes toward black and Hispanic students should be assessed.
4. Interview all students who wish to participate in the program. This time should be used to acquaint them with the ct those mentors whom they perceive as potentially beneficial to their development. Under these circumstances, the role of the coordinator would be to assure an equal distribution of mentors and protégés whenever possible so that all mentors are selected by at least one protégé.

EVALUATION

Successful program development cannot occur without evaluation. From the evaluation priorities among needs may be set and these needs will be translated into program objectives or modifications of existing objectives.

1. program needs assessments
2. individual needs assessment
3. resource allotment
4. processes or strategies for providing mentoring services to students

A structured interview protocol should consist of open-ended questions for a representative sample of students, faculty, administration, and staff. Following derivation, these themes should be supplemented by issues identified by faculty, administration, and staff. The sum of themes should then be converted into Likert-type items. The results will be questions that can be scored and subjected to a parametric statistical analysis. Pilot testing should be performed on a sample population. Following analysis of internal homogeneity, the questions should be readied for group administration.

A confidential inventory should be administered to a random sample of students to determine the precise relationships between student achievement and program activities. Confidentiality will assure candid responses. A random sample will allow extrapolation without the expense of locating all students. The findings and recommendations from this evaluation should provide consultation for strategic planning for the next academic year.

Table 6. Frequency Distribution of Class Level of Selected Participants by Race/Ethnicity

	(N = 99)		
		RaceEthnicity	
	Black	*Hispanic*	*Other*
Freshmen			
N	27	29	0
%	27.2	29.2	9
Sophmores			
N	28	2	0
%	28.2	2.0	0
Juniors			
N	14	1	1
%	14.1	1.0	1.0
Seniors			
N	4	3	0
%	4.0	3.0	9
Graduate Students			
N	0	0	0
%	0	0	0
Undeclared			
N	0	0	0
%	0	0	0
Total			
N	63	35	1
%	63.6	35.3	1.0

In preparation for strategic planning, a feedback session with staff, administrators, and mentors should be convened to discuss the quantitative and qualitative findings.

Table 7. Majors of Selected Student Participants by School

(N = 99)	N
School of Applied Arts & Sciences	26
School of Business	27
School of Engineering	6
School of Humanities & Arts	9
School of Science	7
School of Social Science	12
School of Social Work	3
Undeclared	9

The San Jose State University Faculty Mentor Program is now in its twelfth year of service. Over the years the program has grown to serve more than 300 students with the help of about 130 volunteer faculty and staff mentors and has proven to be a success in enhancing the retention rates of participating students.

REFERENCES

Astin, A.W. (1975). *Preventing students from dropping out.* San Francisco: Jossey-Bass.

Bliss, L.B., & Mueller, R.T. (1987). Assessing study behaviors of college students. *Journal of Developmental Education, II,* 14-18.

Brophy, J.E., & Good, T. (1974). *Teacher-student relationships: Causes and consequences.* New York: Holt, Rinehart and Winston, Inc.

Canton, M.E., & James, D.P. (1994). *Mentoring guide for community colleges* Canton Associates, Supported by the National Science Foundation.

Clewell, B., & Ficklen, P. (1986). Effective institutional practice for inspiring minority retention in higher education. *Journal of College Admissions, 116,* 7-13.

Daloz, L.A. (1987). *Effective teaching and mentoring: Realizing the transformational power of adult learning experiences.* San Francisco: Jossey-Bass.

Pascarella, E.T., & Terenzini, P.T. (1977). Patterns of student-faculty informed interaction beyond the classroom of voluntary freshman attrition. *Journal of Higher Education, 5,* 540-552.

Rowe, M.P. (1981). Building mentorship frameworks as part of an effective equal opportunity ecology. In J. Farley (Ed.), *Sex discrimination in higher education: Strategies for equality.* Ithaca, NY: Cornell University, ILR Press, New York State School of Industrial and Labor Relations.

Tinto, V. (1975, Winter). Dropout from higher education: A theoretical synthesis of recent research. *Review of Educational Research, 65,* 89-125.

Tinto, V. (1987). *Leaving college: Rethinking the causes and cures of student attrition.* Chicago; University of Chicago.

INCREASING GRADUATE STUDIES INTEREST AND ADMISSIONS AMOUNG AFRICAN-AMERICAN UNDERGRADUATES:

A ROLE FOR GRADUATE STUDENTS

Cheryl A. Boyce

INTRODUCTION

The current status of African Americans in graduate studies continues to be distressing and disproportionate. Not until 1995 the did number (1,287) of African-American doctoral receipients surpass the previous high number of 1,116, recorded almost 20 years earlier in 1977 (Henderson, Clark, & Reynolds, 1996). Significantly, from 1977 to 1994 the number of Ph.D.s awarded in the United States to members of other underrepresented ethnic groups increased while for African Americans the numbers showed only decreases compared to 1977 (Ries & Thurgood, 1993; Simmons & Thurgood, 1995).

Diversity in Higher Education, Volume 1, pages 93-102.
Copyright © 1997 by JAI Press Inc.
All rights of reproduction in any form reserved.
ISBN: 0-7623-0086-8

The number of African Americans pursuing graduate degrees may be influenced by the amount of first-generation college students who are African Americans. These students may not consider graduate school even if they have the academic promise. One reason for disinterest in graduate school may be that they have few, if any, role models who have obtained graduate degrees. Another possible reason is that they have faced significant financial and emotional stressors while obtaining their baccalaureate degrees. Before African-American undergraduate students undertake additional years of education, they may need to be convinced that graduate school can provide them with significantly better opportunities.

In most fields a graduate degree will mean long-term career advancement and financial gains. However, it is hard for minorities to give up the relatively high income of a job in order to pursue further study. Graduate school is a more difficult choice for those from poorer families, with children and families of their own, or with large financial obligations from their undergraduate studies (Blackwell, 1981; Solomon & Wingard, 1991). Thus, students need to make a cost-benefit analysis of graduate education.

The average undergraduate student may not have entertained the idea of graduate school before. Even if students want to pursue a graduate degree, they may need certain prerequisite classes and research experiences that are beyond the scope of their baccalaureate degree. Thus, an important part of the bleak picture of African Americans in graduate education is the lack of preparation that occurs while they are undergraduates (Wilson, 1988).

Many African-American students have the academic capabilities to succeed in graduate education, but they need specific knowledge, training, and support to be admitted and to succeed in gradate school. African-American undergraduate students may not have knowledge of the process of applying to graduate schools, of locating funding, and how to select an area of specialization (Blackwell, 1981). The Minority Locator Service of the Graduate Record Examination (GRE) and guides to graduate schools found in local libraries are good resources of information on schools. However, once graduate schools are identified, a student needs to know how to apply competitively.

In addition, some African-American students have negative perceptions about graduate school and research. It has been suggested that there are four reasons why minority students are less likely to participate in research and development: (1) they may lack the mathematical orientation of research and development; (2) they may have perceptions that research has been used against minorities; (3) they may hold the belief that most researchers conduct "pure" research which is abstract and unrelated to the real world; (4) they may be more subject to have desires to do work that will help minorities more directly than seems to be afforded by research (Morrison, 1977).

Thus, the lack of African-American graduate students is related not only to a lack of interest but also a lack of preparation for admission into a reputable graduate school program, completion of their graduate course work, and subsequent success in research.

MENTORING

One procedure that is often used to address the shortage of African-American doctoral recipients is the use of mentoring relationships (Blackwell, 1981, 1983; Frierson, 1981). Mentors have different meanings to developmental psychologists, the business community, and those in academic settings. Most of the definitions of "mentor" follow from the Greek mythological root of the word and suppose that the mentor is an older, more experienced professional person usually of the same sex and race who has an influential position and seniority (Busch, 1985; Kogler-Hill, Bahniuk, Dobos, & Rouner, 1989; Kram, 1985; Lester & Johnson, 1981; Levinson, Darrow, Klein, Levinson, & McKee, 1978). The mentoring relationships are also a form of social support and are purported to have powerful and professional impact on the protégés or students (Bowman, Bowman, & Delucia, 1990). The many diverse definitions of mentor and the mentoring process in the literature, however, may relate to the complexity of the mentor relationship. Perhaps the word "mentor" is inadequate and too narrow to describe this important developing relationship (Levinson et al., 1978). Although the chapters of Charles Healy and Morris Zelditch in this volume provide major contributions in defining the term "mentor" and the reader is urged to review those works.

Despite the definition that is used, the literature on mentoring suggests mentoring is necessary for positive development and growth (Merriam, 1983). In higher education the mentor/student relationship is deemed important for academic success (Cusanovich & Gilliland, 1991). The mentoring process not only provides individuals with support and protection during academic training but also during their professional careers (Frierson, 1990). Mentoring has been found to be highly related to both professional and career development (Roche, 1979; Vaughn, 1985).

Various models of mentoring relationships have been described in the literature. There are models of a "symbiotic" or "clone" student-mentor process (Blackburn, Chapman, & Cameron, 1981), as well as, developmental-contextual frameworks of the mentoring process (Wertsch, 1985). However, an important view of the mentoring process is one which accounts for reciprocity in mentoring as well as discussed by Healy and Welchert (1990) and again by Healy in this volume. According to this model of mentoring the mentoring relationship is a dynamic, reciprocal relationship between a mentor and a student that is aimed at promoting the career development of both. The student's purpose is not only to receive information, but also to transform their identity from an "understudy" to an independent colleague.

A reciprocal relationship is beneficial to the mentor as well. Through a mentor relationship with reciprocity, a mentor can create a new academic and transfer some of their knowledge, identity, and theories to a "protégé" for the future. A mentoring relationship with reciprocity includes a close personal relationship and specific activities such as research grants and publications that further both the mentor's and student's career. These activities that occur in a mentoring relationship with reciprocity would help African Americans get into professional "networks" (Roper, 1980).

MENTORING FOR AFRICAN-AMERICAN STUDENTS

For African-American scholars mentoring has been found to be particularly important for success and completion of graduate programs (Blackwell, 1983). There are many variables that account for an African-American student's success or distress in the pursuit of higher education and research careers. Students are often faced with societal pressures and institutional racism. Without successful mentoring, African-American students are further compromised in their abilities to pursue and succeed in graduate education and to combat the effects of those pressures. The benefits of a true mentoring relationship are clear, yet African Americans tend to be limited in their mentoring relationships (Blackwell, 1983; Frierson, 1990).

One cited barrier for mentoring is the attraction people have to mentors and protégés of the same sex and race (Olian, Carroll, Giannantonio, & Feren, 1988). If this is true and remains the case, there are not enough mentors to go around because of the lack of minority faculty. Consequently, there will never be an increase in the number of African Americans with graduate degrees if we are dependent upon the few minority faculty to be entirely responsible for the future of all the minorities in graduate education.

The barriers to mentoring for African Americans are similar to barriers faced by other minorities in the workplace (Noe, 1988). African Americans may fail to develop mentorships because of limited contact with potential mentors. Moreover, they may not have the knowledge of how to develop informal networks. African Americans may feel like "tokens" which can negatively affect their attitudes, self-perceptions, and skills for graduate studies (Pruitt, 1985). There may also be stereotypes and beliefs about African Americans in graduate education, and negative perceptions about their potential or actual performance. It has been found that some programs which were implemented to aid minorities may have actually stereotyped, undervalued, and stigmatized them (Steele, 1992). Thus, personal relationships with role models such as mentors are necessary to help African Americans deal with their experiences in higher education.

A NEW MODEL FOR MENTORING: GRADUATE STUDENTS AS MENTORS FOR UNDERGRADUATES

Mentoring may be a component that is important for spurring initial interest in graduate studies and a pursuit of a graduate degree. A positive mentoring relationship could benefit the undergraduate protégé's self-esteem and confidence as well as increase his or her interest in graduate school, competitive applications for graduate schools and fellowships, and possibly affect his or her graduate and professional school career. Given the barriers to a true mentoring relationships for most African-American graduate students, there are even more barriers and lack of access to mentors for undergraduate students.

The proposed model presents graduate students as serving in mentor roles to undergraduate students. Although graduate students are not as experienced as faculty, they can still serve as a role model or mentor. A student needs special preparation for success in graduate studies: academic preparation, motivation, information on the graduate school application and admission process, advice on good choices for graduate school, information on financial opportunities, research opportunities. Upon graduation from their baccalaureate institution undergraduate students need survival techniques for graduate education and continuing social support. A graduate student could help an undergraduate student personally with most of these needs or should be able to direct an undergraduate in meeting these needs.

For example, a graduate student mentor may know which classes are prerequisites for their particular field. They may also know which classes will be more useful during graduate study, such as advanced statistics and research courses. Motivation for graduate school has to come largely from the undergraduate protégé but can be fostered and encouraged with support from a graduate student. Since a graduate student has searched for a graduate school more recently, he or she can give first-hand information on where to go for information, which particular schools might meet their interests, as well as students and faculty the undergraduates should contact at those institutions.

Later in the application process a protégé might need a friendly ear and some informal information on the personal and professional impact of a potential school choice. In addition, a graduate student can relate the societal and academic pressures they may face. Some undergraduate students are dismayed by the financial cost of graduate education. Consequently, graduate students may impart knowledge about scholarships, fellowships, and other financial opportunities for graduate school that can encourage and motivate a financially strained undergraduate.

Most importantly, students need specific experiences outside the undergraduate curriculum to become competitive for major graduate school programs. Some programs have been developed to give undergraduates strong foundations in research and development, such as the Minority Access to

Research Careers Program out of the National Institute of General Medical Science, the Ronald E. McNair program out of the Department of Education, the Research Careers for Minority Scientists, formerly out of the National Science Foundation, and other externally sponsored programs. Graduate assistants could have undergraduate students assist them with research for their master's and dissertation theses, thus creating a possible linkage to a faculty mentor—the theses or dissertation director of the graduate student. At some institutions there are already structures in place for a student to gain academic credit for such experiences. Furthermore, through mentoring, a graduate student could gain new ideas and insights into their own research.

Informally, these type of mentoring experiences occur for some undergraduates. Similarly, peer programs have been used for academic achievement for undergraduates and graduate school transitions (Bowman, Bowman, & Delucia, 1990; Jacobi, 1991). New conceptions of mentoring by graduate students could result in the beginnings of an informal network system. The protégés of today become the colleagues of tomorrow. Those who serve as mentor were most likely mentored themselves (Busch, 1985). Such a process where graduate students mentor undergraduate students could be the beginning of a new "old-boy network" for African Americans.

A PROPOSED PROGRAM

A prototype program for graduate student mentors would involve several components. The objectives of the program would be twofold. First, the program would encourage and assist undergraduates in the pursuit of graduate education. A secondary goal would be to provide graduate students with the experience of being a mentor. The important elements of the program would be the advising, counseling, and social support system between the graduate student mentor and the undergraduate student protégé. Specific aspects and purposes of the mentoring program outlined below are taken from the program suggestions from community-based support systems established by the Washington Center, a nonprofit experiential education institution in Washington, DC for the retention of its minority students (Hoard, 1989). The proposed graduate student mentoring program would be implemented so as:

- To help undergraduate students become more motivated to pursue graduate education.
- To help undergraduate students have the necessary skills to succeed in their graduate studies and pursue academia.
- To allow undergraduate students the opportunity to relate to someone similar to themselves who went through a similar experience.

- To help undergraduates relate to and observe the graduate student who has "made it."
- To help undergraduates more effectively develop their communications and network systems.
- The mentors can provide ideas, issues, alternatives, advice, and guidance to the undergraduate student.
- To help both undergraduate and graduate students develop more positive attitudes about themselves.
- The relationship will result in positive benefit to the undergraduate students and they will in turn mentor and motivate someone else.

Both the mentor and protégé would need training, information, and support from faculty and staff. Needless to say, the program should be careful to avoid stereotypes or stigmas that African Americans are not prepared for graduate school. On the contrary, the participating undergraduate students should be reminded that they are capable of getting into a graduate school and may need additional resources that are available to many other students.

The training for the graduate student mentors should provide structured information on mentoring, communication skills, and resources for them and their undergraduate protégé. Ideally, the graduate student's own advisor or mentor should be involved in a research component structured for academic credit for the undergraduate. A stipend for the undergraduate and graduate could provide added motivation as well as possibly needed financial support.

Matches between the graduate student mentor and undergraduate student protégé should be based on the students' stated interests. The program should begin for the undergraduate student when he or she starts his or her junior year, at a time when a student usually chooses a major field of study. The graduate student mentors should be at least in their second year. By the second year graduate students have become more settled in their graduate program and would have more time to give to an undergraduate. As the graduate students' develop, both academically and in their research skills, so can their mentoring relationship from year to year.

A positive research experience through this program model might include completion of a scholarly paper or honors thesis by the undergraduate protégé. According to Healy's model of mentoring which includes reciprocity, the undergraduate student and graduate student mentor's perceptions would not only be expected to be positive, but it is also expected that the mentor and student will have a closer personal relationship that includes the formation of a student's career and identity. Thus, the long-term goal of the program would be to create a mentoring relationship with reciprocity which would strengthen a graduate school applicant and prepare the undergraduate personally and academically for graduate studies.

Evaluations of this program should occur twice a semester, at the beginning and the end. In-depth interviews between graduate student mentors and the undergraduate protégés should be conducted that address the relationships between mentors and protégés. The interviews should specifically address the impact of the mentor relationship on the student's research experience as well as their interest in the pursuit of graduate study should be examined. Near the end of the last semester for the undergraduate student protégé, information on the student's choice to pursue graduate studies, type and number of schools applied, and admissions should be documented. Yearly follow-up data on the protégé should be conducted to assess the long-term impact and usefulness of the mentor relationship for the undergraduate.

Although the program is designed to help undergraduates, the mentor benefits from his or her role. Therefore, the mentor's experience should be particularly interesting to document. At the end of the formal mentoring period when a protégé finishes his or her baccalaureate degree, the mentor may have experienced personal growth and development through the mentoring experience. Such a positive experience could enhance their graduate career.

CONCLUSIONS AND RECOMMENDATIONS

The intent of this paper was to introduce serious thought regarding the role graduate students can take in the development of prospective graduate students. Given the lack of minority faculty, minority graduate students may be a good substitute for the role that a minority faculty member might have in the development of new African-American graduate students.

There are many benefits to the mentor, protégé, and network of African-American students through this model. Both the mentor and protégé should have higher self-esteem as a result of such experiences. If a research component is in place in the mentor-protégé relationship, the mentor would not only have a research assistant, but a source of new ideas and teaching experience. The mentor would benefit from the positive feelings that would come from creating a protégé, a new colleague. The protégé would get invaluable information and preparation about graduate school. The undergraduate protégé would benefit from social support, advice, and knowledge that should continue through their graduate school years and beyond.

The so-called "old boy" network has served many traditional graduate students well. African Americans and others who are significantly underrepresented need to develop their own networks. Networks contribute to professional advancement, disseminate information, and most importantly provide social support.

With the disproportionate amount of African-American faculty, African-American graduate students are an untapped resource who need positive experiences such as being a mentor. From my own experiences as a recent graduate student, I have found that informal and formal mentoring of undergraduate students resulted in increased self-esteem, higher confidence, new research ideas, and a new network of future colleagues.

It is important to begin to carry out programs to facilitate mentoring among undergraduate students to encourage them to pursue graduate studies, help them gain admission, and then succeed in their graduate studies. Current research programs that are designed to encourage minorities in research careers are in place and may foster faculty mentor relationships. However, these programs are not enough to reach the many promising undergraduate African-American students and may lack the personal social support and informal information that a graduate student can provide.

Implementating a mentoring program using a model with graduate student mentors could result in a positive rippling effect and network of African-American graduate students who will become professors and other professionals. Until then, it is the responsibility of graduate students to reach out and mentor others as much as they can to begin the network. The benefits of mentoring are immeasurable for the mentors and protégés. The supportive network system that mentoring builds is needed to increase the presence of African American's in graduate school and academia.

REFERENCES

Blackburn, R.T., Chapman, D.W., & Cameron, S.M. (1981). "Cloning" in academe: Mentorship and academic careers. *Research in Higher Education, 15* (4), 315-327.

Blackwell, J.E. (1981). *Mainstreaming outsiders: The production of black professionals*. Bayside, NY: General Hall, Inc.

Blackwell, J.E. (1983). *Networking and mentoring: A study of cross-generational experiences of blacks in graduate and professional schools*. Atlanta: The Southern Educational Foundation.

Bowman, R., Bowman, V., & Delucia, J. (1990). Mentoring in a graduate counseling program: Students helping students. *Counselor Education and Supervision, 30*, 58-65.

Busch, J.W. (1985, Summer). Mentoring in graduate schools of education: Mentors' perceptions. *American Educational Research Journal, 22* (2), 257-265.

Cusanovich, M., & Gilliland, M. (1991). Mentoring: the faculty-graduate student relationship. *Council on Graduate School Communicator, 24*, 1-2.

Frierson, H.T. (1981). Minority participation in R&D: Developing an undergraduate feeder system. *Journal of Negro Education, 50*, 401-406.

Frierson, H.T. (1990, March). The situation of black educational researchers: Continuation of a crisis. *Educational Researcher, 19*, 12-17.

Healy, C.C., & Welchert, A.J. (1990, December). Mentoring relations: A definition to advance research and practice. *Educational Researcher, 19*, 17-21.

Henderson, P.H., Clarke, J.E., & Reynolds, M.A. (1996). *Summary report 1995: Doctorate recipients from United States universities*. Washington, DC: National Academy Press.

Hoard, D.W. (1989). Models of community resources for the enhancement of black student retention. In J.N. Niba & R. Norman (Eds.), *Recruitment and retention of black student in higher education* (pp. 71-88). Lanham, MD: University Press of America.

Jacobi, M. (1991, Winter). Mentoring and undergraduate academic success: A literature review. *Review of Educational Research, 61* (4), 505-532.

Kogler-Hill, S.E., Bahniuk, M.H., Dobos, J., & Rouner, D. (1989). Mentoring and other communication support in the academic setting. *Group and Organization Studies, 14*, 355-368.

Kram, K.E. (1985). *Mentoring at work: Developmental relationships in organizational life.* Glenview, IL: Scott, Foresman & Co.

Lester, V., & Johnson, C. (1981). The learning dialogue. Mentoring. In J. Fried (Ed.), *Education for student development. New directions for student services: No. 15* (pp. 49-56). San Francisco: Josey-Bass.

Levinson, D.J., Darrow, C.M., Klein, E.G., Levinson, M.H., & McKee, B. (1978). *The season's of a man's life.* New York: A.A. Knopf.

Merriam, S. Thomas, T., & Zeph, C. (1987). Mentoring in higher education: What we now. *Journal of Higher Education, 11*, 199-210.

Morrison, B.M. (1977, April). *Training minority researchers for academia.* Paper presented at the American Educational Research Association Annual Meeting.

Noe, R. (1988). Women and mentoring: A review and research agenda. *Academy of Management Review, 13* (1), 65-78.

Olian, J.D., Carroll, S.J., Giannantonio, C.M., & Feren, D.B. (1988). What do proteges look for in a mentor? Results of three experimental studies. *Journal of Vocational Behavior, 33*, 15-37.

Pruitt, A.S. (1985). Discrimination in recruitment, admission, and retention of minority graduate students. *Journal of Negro Education, 54* (4), 526-537.

Ries, P., & Thurgood, D.H. (1993). *Summary report 1991: Doctorate recipients from United States universities.* Washington, DC: National Academy Press.

Roche, G.R. (1979), Much ado about mentors. *Harvard Business Review, 57*, 24-27.

Roper, D. (1980). The waning of the old boy network: Placement, publishing, and faculty selection. *Improving College and University Teaching, 28*, 12-18.

Simmons, R.O., & Thurgood, D.H. (1995). *Summary report 1994: Doctorate recipients from United States universities.* Washington, DC: National Academy Press.

Solomon, L.C., & Wingard, T.L. (1991). The changing demographics: Problems and opportunities. In P.G. Altbach & K. Lomotey (Eds.), *The racial crisis in American higher education* (pp. 19-42). Albany, NY: State University of New York Press.

Steele, C.M. (1992). Race and schooling of black Americans. *The Atlantic Monthly*, 68-78.

Vaughn, J.R. (1985). Minority students in graduate education. In B. Smith (Ed.), *The state of graduate education.* Washington, DC: Brookings Institution.

Wertsch, J.V. (1985). *Culture, communication, and cognition: Vygotskian perspectives.* New York: Cambridge University Press.

Wilson, R. (1988). Developing leadership: Blacks in graduate and professional schools. *Journal of Black Studies, 19* (2), 163-173.

THE RECRUITMENT AND RETENTION OF TALENTED AFRICAN AMERICANS IN SCIENCE:

THE ROLE OF MENTORING

Charles Woolston, Freeman A. Hrabrowski, III, and Kenneth I. Maton

INTRODUCTION

The Meyerhoff Scholarship Program was created at the University of Maryland Baltimore County (UMBC) in 1988 with a half-million dollar grant from the Robert and Jane Meyerhoff Foundation. The grant was initially designed to increase the number of African Americans, particularly males, who successfully pursue careers in science and engineering. The first grant enabled UMBC to launch a program of full support for outstanding African-American students, competitively selected from across Maryland, who would major in mathematics, a science, or a technical discipline and then pursue terminal degrees in these areas (emphasizing Ph.D.s and M.D./Ph.D.s). The first group

Diversity in Higher Education, Volume 1, pages 103-114.
Copyright © 1997 by JAI Press Inc.
All rights of reproduction in any form reserved.
ISBN: 0-7623-0086-8

of Meyerhoff scholars included 19 young men who enrolled in fall 1989. The program now has a total enrollment of 192 students from across the nation (93 males, 99 females), including 45 new Meyerhoff freshmen in fall 1997. The program is now open to all high-achieving students who have an interest in pursuing doctoral degrees in the sciences or engineering, and who are interested in the advancement of minorities in these fields.

The program enjoys support from a variety of national agencies and corporate sponsors, in addition to the Meyerhoff Foundation. Program participants represent one of the largest concentrations of high-achieving African-American students majoring in science and engineering anywhere in the United States.

Many indicators point to the success of the Meyerhoff Program and its students. Examples are listed below.

1. Exceptional graduation rates (94%) in science and a cumulative mean grade-point average of 3.4.
2. Impressive research internships in laboratories throughout the United States and abroad, resulting in numerous presentations at national conferences and several articles in refereed journals.
3. Postgraduate work by students in the program's first four graduating classes (1993-1996): 98 students have graduated from the program. Currently, 80 are enrolled in graduate or professional schools in science, engineering, medicine, or related areas at nationally recognized universities. (The majority of the other graduates intend to enter graduate school in the near future.) Examples of these programs include MIT's Ph.D. programs in Physics and Toxicology, Cornell University's Ph.D. program in Organic Chemistry, the University of Michigan's and Northwestern University's Ph.D. Programs in Electrical Engineering, the University of North Carolina's Ph.D. program in Chemistry, the University of Virginia's Ph.D. program in Environmental Sciences, Rice University's, and the University of Maryland College Park's Ph.D. programs in Mathematics, the University of Florida-Gainesville's Ph.D. program in Computer Science, and M.D./Ph.D. programs at the University of Pennsylvania, Johns Hopkins University, and the University of Maryland at Baltimore, and M.D. Programs at Harvard, Yale, and Duke.

THEORETICAL BASE

Although African-American students represent 9 percent of those enrolled in universities and colleges, they earn only 5 percent of the bachelor's degrees, and approximately 2 percent of the doctorates in science and engineering. The fact that most talented African-American students do not earn degrees in their initially chosen major in the sciences is of particular concern not only for these youths, but for higher education and the nation as a whole. The research literature suggests that the students' lack of success is due to an array of factors (e.g., poor academic preparation, lack of adequate academic and social support systems). Fortunately, the research literature also suggests a number of factors that are critical for minority student success, performance, and persistence in science.

Specifically, the literature suggests that the following factors are critical for minority student success in terms of performance and persistence in science: (1) knowledge and skills; (2) motivation and support; (3) monitoring and advising; and (4) academic and social integration. The Meyerhoff Scholarship Program focuses explicitly on these factors and relies on five major sources of influence on students: peers, faculty, administrative staff, family, and community members, including Baltimore-Washington area African-American professionals in science, engineering, and medicine. Each of these groups can provide "mentors" who can increase the chances of success for minority students.

The program is based on a model, which assumes that every student selected has the ability to succeed in science and engineering, given the appropriate opportunities and resources. The Meyerhoff Scholarship Program consists of 13 components which, together, help to create an environment that continually challenges and supports students. Unlike most intervention programs discussed in the literature, ranging from two-weeks to six-weeks and lasting up to two academic years, the Meyerhoff Scholarship Program provides substantial support for students from their pre-freshman summer through graduation and beyond. In addition, the program's emphasis on having a diversity of mentors provides various forms of support across a number of disciplines.

The theoretical components consist of the following:

1. *Recruitment* of top high school math and science students and an on-campus selection weekend involving faculty, staff, and peers.
2. *A Summer Bridge Program* including math, science, and humanities course work, training in analytic problem solving, group study, and social and cultural events.
3. *Scholarship support* for Meyerhoff scholars (four-year comprehensive scholarships) and for Meyerhoff finalists (limited support); continued support is contingent on maintaining a B average and a science or engineering major.
4. *Study groups.* Group studying among students is strongly and consistently encouraged by program staff.
5. *Program values* consistently emphasized include striving for outstanding academic achievement, seeking help (tutoring, advisement

counseling) from a variety of sources, supporting one's peers, and preparing for graduate or professional school.

6. *Program community.* The Meyerhoff Program represents a family-like, campus-based community for students. Staff regularly conduct group meetings with students; students live in the same residence halls during their freshman year.
7. *Personal advising and counseling.* A full-time academic advisor, along with the program's executive director and other staff, regularly monitor and advise students.
8. *Tutoring.* All Meyerhoff students are encouraged to take advantage of departmental and university tutoring resources in order to optimize course performance.
9. *Summer research internships.* Program staff use a network of contacts to assist students in arranging summer science and engineering internships during the summers.
10. *Faculty involvement.* Department chairs and faculty are involved in all aspects of the program, including recruitment, teaching, research mentorship, and special events and activities.
11. *Administrative involvement and public support.* The Meyerhoff Program receives high-level campus administrative support, and high-level public support.
12. *Mentors* for each student are recruited from among Baltimore-Washington-area professionals in science, engineering, and health.
13. *Family involvement.* Parents are kept informed of student progress, invited to special counseling sessions if problems emerge, included in various specials events, and have formed a mutual support resource, the Meyerhoff Family Association.

These components, taken together, address each of the four primary areas of minority student need reviewed earlier. Most important, they do so by mobilizing and building upon sources of influence mentioned earlier. While science-related mentors are important enough to warrant a separate listing as one of the primary components, the concept of mentoring is at the heart of almost every component. The success of the program assumes that minority students, to be successful in the study of science and math, benefit enormously from support from every segment of the campus. Mentors, both formal and informal ones, therefore are positioned at every stage of the program to be supportive of the students.

MENTORING ELEMENTS

The structure of the Meyerhoff Scholarship Program resembles that of a family unit, and the students look to their mentors to provide support and advice much as they look to their parents. The program arranges a "family meeting" at the beginning and end of each semester for the entire group of students. Program staff conduct additional meetings for each class of Meyerhoff students once per semester, and for other small groups as needed. All students are expected to attend meetings and to talk about their experiences, including their successes and failures. Achievement is not only acknowledged but celebrated. Straight-A averages are rewarded with standing ovations by the students. "A" grades on exams in particularly difficult courses are also rewarded with ovations. The intent of these meetings is to build pride in academic achievement and to give support to those students experiencing difficulty.

Students at family meetings are also encouraged to share their disappointments. Sometimes this develops into a discussion of a problem with a particular faculty member. Upper-class students are asked to offer advice on solving the problems. The students sharing the problems are often asked about their study habits and whether they are seeking help from other students, either within or outside the Meyerhoff Program. This exercise is aimed at making students feel comfortable about raising issues or problems, sharing difficulties, and seeking help. In this way the students begin to take responsibility for the success of their fellow Meyerhoff students. Most important, students become comfortable analyzing their studying effectiveness and academic performance.

One cannot overemphasize the role of the senior staff as mentors to all the Meyerhoff students. They provide direction for the program and for individual students in the program. Their relationships with the students, the respect they command on the campus, and their academic advising expertise, especially in science and engineering, provide them with an unusual opportunity to diagnose problems and to suggest solutions. The other mentor relationships within the Meyerhoff Program supplement the mentoring provided by senior staff.

Formal Mentors

The most formal of all the mentors in the Meyerhoff Program are those from the larger scientific, engineering, and medical communities. These mentors are identified by the staff through personal contacts. They are successful African-American professionals from the Baltimore-Washington area who are willing to devote time interacting with Meyerhoff students during their undergraduate years. They are expected to attend the mentors meetings and to meet with their mentees at least twice each semester. Contact with a successful professional from the scientific community is very valuable for the students. At the very least it gives the student a perspective on the scientific work world. In the best relationships the mentor becomes an excellent role model and friend who can expose students to practical applications of science.

One challenge is to correct the reality that some of the mentors fail to follow through on promises to spend time with students for whom they were to serve as mentors.

Faculty

In addition to working with science mentors, Meyerhoff students are encouraged to develop strong relationships with faculty members within and outside their academic disciplines. They are encouraged to go beyond the conventional advisor/advisee connection. The students are encouraged to participate in the faculty members' research activities whenever feasible. This not only gives the student a connection to the department, but also involves them directly in research projects. A number of students work in faculty labs beginning in their freshman years. These research experiences are especially beneficial to the students when they apply to graduate schools. They have had substantive research experiences, and the faculty they have worked with can comment first-hand on their abilities in letters of recommendation. In many cases, UMBC faculty contact colleagues to place students in summer research labs and later into graduate schools across the nation.

Students' close interactions with faculty have yielded several interesting results. Faculty have high expectations of Meyerhoff students and monitor their performance carefully. This often results in faculty calling program staff to report either downward trends in students' performances or examples of exceptional achievements in the course. This allows the staff to intervene directly to determine the cause of students' problems when they arise, and to work with students to address them, or to highlight excellence. Sometimes the staff helps the students simply by assigning them to a study group or finding them a tutor; in other cases staff may determine that students need counseling support to handle personal problems. The important element is that faculty and program staff interact regularly.

Staff

The professional staff, in addition to handling administrative responsibilities, all serve as mentors for the Meyerhoff students. They serve as academic advisors, confidants on personal issues, conduits to the faculty, and a source of information for campus services. They are "the heart" of the support program. Currently, five professional staff members work with the 163 Meyerhoff students.

It is helpful to describe the responsibilities of one of the professional staff as an example of the staff's role in ensuring the academic success of students. This staff member coordinates student placements in summer internships and graduate schools, and works with other campus offices that have similar responsibilities. Specifically, this staff member works with universities and industry to identify research opportunities for students throughout the scientific community, both nationally and abroad. Students have been placed with Nobel Prize-winning scientists, and they have been placed at universities in other countries and in some of the top industrial research labs in the nation—in addition to labs on the UMBC campus. These opportunities expose students to situations where they can observe and participate in truly cutting-edge research. In many cases these students emerge from these experiences with new "mentors" who can help to guide them through the "ins" and "outs" of the scientific community. Supporting students as they apply to graduate and professional schools requires considerable counseling and advising. As needed, this staff member encourages students to adhere to deadlines, prepare for standardized testing, and secure the most impressive recommendations. Every student meets with this staff member during the fall of his or her junior year and begins to prepare a list of potential graduate or professional schools. Students are responsible for gathering information on those schools and preparing a plan for the application process. In addition, the staff member maintains records on each student and regularly monitors students' progress during the application and admission process.

The "family" concept, which represents the basis of the Meyerhoff support system, is most evident in the relationship of the staff and the students. Staff members work with the Meyerhoff students at all levels of the undergraduate experience. They work with them individually to solve personal and academic problems, to negotiate the university bureaucracy, and to learn to monitor their own academic progress. As a group they work with them to help them learn how both to provide and seek help, to form study groups, and to use the group's strengths to find academic support.

The strong academic support system includes closely monitoring academic progress. Students are expected to report difficulties as soon as they become obvious. If students fall below a B average (3.0) overall or in their major, they are placed on Meyerhoff Program probation. Individual "recovery" plans are devised and agreed upon by the staff and the student. Most often, such a plan involves a weekly meeting between the staff member and the student. Faculty members are often partners in the plan and provide feedback to the staff on a regular basis. Fortunately, because very few students ever encounter this kind of difficulty, the case load is relatively low (ranging from 10 to 20%), and faculty and staff can work closely with individual students.

Peers

The Summer Bridge Program provides the first opportunity for peer mentoring. Older Meyerhoff students serve as resident counselors and many also are on campus during the summers conducting research. They are especially helpful to the new students, offering insights about the campus and the program. While the freshmen are bonding into a cohort group, the presence of the older Meyerhoff students provides the mechanism for mentor/mentee relationships to develop.

The family atmosphere also encourages the students to offer and seek help among their peers. These relationships develop through the study groups that form, the voluntary tutoring that occurs, and the family and peer group meetings that are held throughout the year. Although there is no formal assignment of older students to mentor the younger Meyerhoff students, the staff strongly encourages these relationships.

Family Support

There is a strong Meyerhoff Family Association. This group focuses on helping the families to support the students in dealing with the university and the scientific community. They raise money for supplementary scholarships and for special activities. One of the Family Association's most important contributions is working with the program staff to assist them in providing support the students need during the difficult adjustment period.

The individual families also are an important component of the program's support structure. The staff works with individual families, often talking with the parents before deciding on an appropriate strategy when a student needs help. This not only keeps the family involved in the student's educational experience, but it also provides the staff with input from the primary support unit. The families have been so important in the student's success that we completed a book on the role of parenting in ensuring academic achievement (Hrabrowski, Maton, & Grief, in press).

Results

The postgraduate data from the program are very impressive. Almost 82 percent (81.6%—80 of 98) of Meyerhoff graduates are currently attending graduate or professional school. Among those in graduate school, 73.8 percent are in science or related areas, and 26.2 percent are in medical school.

One of the program's most distinctive characteristic is its evaluation component, which was included in the program's original design. The founder believed strongly that it was essential to be able to demonstrate its success using the research tools of the academy. Therefore, the Meyerhoff Scholarship Program has a variety of research findings, most of which illuminate the program's success. Outcome analysis indicates that the first four Meyerhoff cohorts were substantially more likely to persist in science and engineering majors than two African-American comparison samples of comparable academic ability. In addition, they achieved significantly higher grades in their science and engineering course work (see Tables 1 through 4).

Table 1. Science vs. Non-science Major after Three Years of College: UMBC Historical (Pre-Meyerhoff) and Current Comparison Samples and Meyerhoff Students

Historical Samples	*Science Major*	*Non-science Major*	*Total N*
African-American	30 (56.6%)	23 (43.4%)	53
Asian	106 (65.8%)	55 (34.2%)	161
Caucasian	619 (56.3%)	480 (43.7%)	1099
Current Samples			
Meyerhoff (African American)	105 (93.8%)	7 (6.3%)	112
Asian	85 (73.9%)	30 (26.1%)	115
Caucasian	237 (51.5%)	223 (48.5%)	460

Note: The chi-square analysis was statistically significant, $X2(5)=83.7$, $p<.001$.

The first report from this program (Hrabowski andMaton, 1995) presented first-year outcomes for the program, comparing the performance of the first three cohorts of students with those of a historical African-American cohort. The data indicated clearly that the Meyerhoff students achieved at a substantially higher level than comparison students, especially in key freshman-year math, science, and engineering courses. These differences were striking in magnitude. Furthermore, the relatively mediocre performance of the comparison sample indicated that, lacking a multifaceted support program, even very talented African-American students are generally unlikely to earn Ph.D.s in the sciences, and, thus, the nation would be unlikely to achieve its goal of substantially increasing the number of Ph.D.s awarded to African Americans in science.

The latest research (Maton & Hrabowski, 1997) incorporates multiple comparison samples and examines outcomes after four years of college for the first four cohorts of students. Meyerhoff students were compared to (1) non-Meyerhoff African-American, Asian, and Caucasian science students of comparable academic capability who attended UMBC before the inception of the program, and Asian and Caucasian science students of comparable academic capability who began attending UMBC since the program began; and (2) students who were accepted into the Meyerhoff Program, but who declined the offer and attended another university instead.

Table 2. Science GPA[1] after Three Years of College for Students Still Majoring in Science: UMBC Historical (Pre-Meyerhoff) and Current Comparison Samples and Meyerhoff Students

Historical Samples[3]	*N*	*Actual Mean*	*Adjusted*[2] *Mean*	*Standard Deviation*
African-American	30	2.5	2.6	0.6
Asian	100	3.0	3.0	0.7
Caucasian	615	3.0	2.9	0.6
Current Samples[4]				
Meyerhoff (African-American)	105	3.1	3.1	0.5
Asian	83	3.1	3.0	0.6
Caucasian	231	3.1	3.0	0.6

Notes: The MANCOVA Ethnicity *X* Time Period interaction was significant, $F(2,1152) = 4.2, p < .05$. Covariates were SAT-Math, SAT-Total, high school GPA, and gender.

[1] All biology, chemistry, computer science, engineering, mathematics, and physics courses taken during the first three years of college were included in calculating the science GPA.

[2] Means were adjusted for the covariates listed above.

[3] Historical samples entered UMBC prior to the start of the Meyerhoff Program (1989), met the same minimal academic criteria as students accepted in the Meyerhoff Program (generally, SAT-Math above 550, SAT-Total above 1050, and high school GPA above 3.0), and had a declared science major while at UMBC or had a freshmen year science course load which included key math, science, or engineering courses.

[4] Current Asian and Caucasian samples entered UMBC since the start of the Meyerhoff Program (1989), met the same minimal academic criteria as students accepted in the Meyerhoff Program (generally, SAT-Math above 550, SAT-Total above 1050, and high school GPA above 3.0), and had a declared science major while at UMBC or had a freshmen year science course load which included key math, science, or engineering courses. The Meyerhoff sample includes the first four cohorts of students (entered UMBC between 1989 and 1992).

The vast majority of Meyerhoff students (93.8%) maintained a science major through three years of college, with all of the comparison samples much less likely to do so (ranging from 56.6 to 73.0%). Furthermore, whereas the four-year science GPA of pre-Meyerhoff UMBC African-American students who retained a science major was below that of their Asian and Caucasian counterparts, the Meyerhoff students who retained a science major reversed this pattern and performed at a level slightly higher than their Asian and Caucasian counterparts.

There were similar patterns of findings when Meyerhoff students were compared to African-American students who were accepted into the Meyerhoff program but declined the offer and attended another university.

It will be a number of years before we can examine the primary criterion for judging the Meyerhoff Program's success, that is, students obtaining a doctoral or professional degree in the sciences or engineering. Nonetheless, the findings to date clearly indicate that the program is having a measurable and

Table 3. Science vs. Non-science Current Major: Meyerhoff Students and Students Who Declined Meyerhoff Scholarship Offers (and attended another University)[1]

	Non-science Majors	*Science Majors*	*Total N*
Meyerhoff Students (Year of UMBC Entry 1989-1992)	105 (93.8%)	7 (6.3%)	112
Students who Declined Meyerhoff Offer (Year of College Entry 1989-1992)	11 (55.0%)	10 (45.0%)	21

Note: The chi-square analysis was statistically significant, X2 (1) = 28.9, $p < .001$.

dramatically positive impact. Meyerhoff students are much more likely to retain a science, engineering, or mathematics major than UMBC comparison samples, and most are going on to graduate schools in the sciences (primarily Ph.D. programs). Furthermore, whereas African-American science students attending UMBC prior to the inception of the Meyerhoff Program achieved science GPAs lower than their Asian and Caucasian counterparts, the African-American Meyerhoff students are performing at a level slightly higher than

Table 4. Science GPA[1] Freshmen Year of College: Meyerhoff Students and Students Who Declined Meyerhoff Scholarship Offers (and Attended Another University)[2]

	N	*Actual Mean*	*Adjusted[3] Mean*	*Standard Deviation*
Meyerhoff Students (Year of UMBC Entry 1989-1992)	112	3.5	3.5	0.4
Students who Declined Meyerhoff Offer (Year of College Entry 1989-1992)	20	2.8	2.8	0.6

Notes: The difference between groups was statistically significant, $F(1,124) = 37.7, p < .001$.

[1] All biology, chemistry, computer science, engineering, mathematics, and physics courses taken during the first year of college were included in calculating the Science GPA.

[2] Four students who did not take science-oriented course work their freshmen year, and one student whose university does not provide freshmen year grades were excluded from the sample.

[3] Means were adjusted for the following covariates: SAT-Math, SAT-Total, high school GPA, gender, year of college entry, and declared college major (science vs. non-science).

their Asian and Caucasian counterparts. Of special import, comparable retention in science, and science GPA patterns, was found when comparing Meyerhoff students to a second comparison sample—students who were accepted into the Meyerhoff Program but decided to attend another university.

As part of an ongoing evaluation process for the Meyerhoff Program, questionnaire and observational data indicate that program-based study groups, peer-based community, the Summer Bridge Program, and program staff—all containing a substantial mentoring component—are especially significant features of the program's success. It appears possible to enhance dramatically the numbers of African-American (and other minority) college students who succeed in science, engineering, and mathematics, given the resources and commitment to do so on the part of our nation and our universities.

CONCLUSION

Undoubtedly, no component of the Meyerhoff Program is more critical than mentoring. What is clear, however, is that mentoring occurs through both formal and informal approaches. What makes this component so effective is that mentoring is at the core of the program's basic philosophy—of those to whom much is given much is expected. Meyerhoff students have the benefit of mentoring from older students, staff, and science and engineering faculty, practicing scientists, engineers, and physicians, and members of the family association. Equally important, students are taught the importance of seeking help and listening to advice provided by others, and mentors are strongly encouraged to be protective in supporting students. The result has been and continues to be an outstanding success story—large number of African Americans excelling in science.

REFERENCES

Hrabowski III, F.A., Maton, K.I., & Greif, G.L. (in press). *Beating the odds: Raising academically successful African-American males.* New York: Oxford University Press.

Maton, K.I., & Hrabowski III, F.A. (1997). *The Meyerhoff program: Fifth year outcomes.* Manuscript in Preparation.

MENTORING AS A TOOL FOR INCREASING MINORITY STUDENT PARTICIPATION IN SCIENCE, MATHEMATICS, ENGINEERING, AND TECHNOLOGY UNDERGRADUATE AND GRADUATE PROGRAMS

William McHenry

OVERVIEW

The Workforce

Twenty-first century economies will be based on how well nations use knowledge derived from science, mathematics, engineering, and technology (SMET) research. A nation's use of this knowledge is dependent on the quality of the SMET workforce and the SMET literacy of its citizens. The United States and other nations must continuously strive to improve their SMET knowledge base by ensuring that all students have access to a quality education system, from kindergarten through adult continuing education programs.

Diversity in Higher Education, Volume 1, pages 115-140.
Copyright © 1997 by JAI Press Inc.
All rights of reproduction in any form reserved.
ISBN: 0-7623-0086-8

Currently, individuals from certain racial/ethnic groups are underrepresented in all areas of the SMET workforce and are underserved at all education levels that prepare individuals for careers in SMET. For example, blacks make up 12 percent of the U.S. population, yet they received 6 percent of the associate degrees, 5 percent of the bachelor's degrees, 2 percent of the master's degrees, and only 1 percent of the doctoral degrees in SMET disciplines in 1995. To remain competitive globally, the United States must provide better access to quality SMET programs for all students, especially blacks, Hispanics, Native Americans, and Pacific Islanders.

Diversity in the SMET workforce is important for many reasons as the examples below indicate:

- It provides the nation with the number of scientists, engineers, mathematicians, and technologists needed for a globally competitive workforce.
- It provides the nation with problem solvers who have the diverse views required to address comprehensive solutions to modern problems.
- It ensures that all groups in society are represented when decisions are made regarding the influence of scientific and technological advances.

The National Science Foundation's (NSF) Research Careers for Minority Scholars (RCMS) program was initiated to encourage individuals from groups underrepresented in SMET disciplines to complete undergraduate degree programs and matriculate to SMET graduate degree programs. RCMS made its first awards in fiscal year 1989. The RCMS program reaffirmed that minority individuals can achieve in SMET disciplines if: financial barriers are reduced; early linkage between classroom knowledge and research experiences occurs; adequate program advising takes place; and students are linked with faculty advocates, preferably faculty mentors.

While the RCMS program encouraged the establishment of strategies that dealt with each of the above, for this chapter the focus will be on lessons learned in creating a supportive mentoring environment.

THE RCMS RECORD

Between 1989 and 1994 RCMS project directors and faculty members from 53 RCMS projects served as advisors for more than 2,300 minority undergraduate scholars who had majors in mathematics, engineering, and technology fields of study.

- More than 57 percent of these students were black, 29 percent were Hispanic, and 11 percent were Native Americans; about 45 percent were women.

- Scholars in the program studied a variety of scientific and technological disciplines, including engineering (37%), biology (13%), physics (12%), chemistry (13%), mathematics (9%), and computer science (6%).
- By 1994 more than 500 RCMS scholars had graduated with SMET degrees. Over half of the graduates (57%) have entered graduate school. Many of the other RCMS graduates considered, or were in the process of, applying to graduate school programs.

RCMS encourages colleges and universities to establish supportive mentoring environments. A key aspect of the RCMS program is that it pairs students with faculty mentors, often eminent scientists and engineers, who guide students' academic and career development. Mentors are vital because they not only inspire but assist students in setting and reaching realistic goals. Mentors can serve to broaden students' horizons, assist them in locating university resources, and combat student isolation. This chapter discusses faculty and student roles and responsibilities in mentoring relationships.

INTRODUCTION

Competitiveness in a High-Technology Global Market

In August 1994 President William J. Clinton and Vice President Albert Gore, Jr. said, "Technology, the engine of economic growth, creates jobs, builds new industries, and improves our standard of living. Science fuels technology's engine. It is essential to our children's future that we continue to invest in fundamental research. Equally important, science and mathematics education must provide our children with the knowledge and skills they need to prepare for high-technology jobs of the future, to become leaders in scientific research, and to exercise the responsibilities of citizenship in the twenty-first century."

Twenty-first century economies will be based on how well nations use knowledge as a resource, especially knowledge produced from science, mathematics, engineering, and technology (SMET) research. Jobs, economic growth, community health, societal stability, and effective government are directly dependent on the health of the SMET workforce and the SMET literacy of the nation's citizenry. In order for the United States to remain competitive in the global markets of the twenty-first century, America must maintain a healthy SMET workforce.

The vitality of the SMET workforce depends on establishing an education system that provides better access for all students to quality SMET programs. Changing demographics mean that strategies must be implemented to recruit students from minority groups traditionally underrepresented in SMET fields to the SMET workforce. Diversity in SMET disciplines is vital because science depends on critical thinking about comprehensive solutions to problems. Creative and innovative solutions arising from new points of view and coming from different worldviews have the potential to further broaden scientific and technological horizons.

Also, diversity in the SMET workforce plays an important part in strengthening society. As individuals learn to work together to solve problems presented by natural disasters, environmental challenges, new and existing diseases, and areas of modern technological development, individuals begin to identify common needs and desires and function as a cohesive society.

Currently, blacks, Hispanics, Native Americans, and Pacific Islanders are underrepresented in SMET disciplines, as evidenced by the relatively small percentages of individuals from minority groups who complete baccalaureate degrees in SMET fields. In 1993 although 281 of the nation's 3,600 postsecondary colleges and universities awarded science and engineering doctorates, about half of these institutions did not grant such a doctorate to even one black student, and only 55 percent of these institutions awarded a science and engineering doctorate to a Hispanic student. This must change if the SMET workforce is to serve the needs of our nation.

The National Science Foundation's (NSF) Research Careers for Minority Scholars (RCMS) program was designed to encourage individuals from minority groups currently underrepresented in the SMET workforce to complete undergraduate SMET degrees. The program also helps individuals who have exhibited unusual talent in SMET disciplines to continue their education and enroll in SMET graduate programs. The program has been a catalyst for positive institutional changes in the culture of SMET disciplines.

The RCMS program serves as evidence that universities and industries working in partnership with the federal, state, and local governments can increase the diversity of students who matriculate through our world-class education system and thereby meet the SMET workforce needs of the country.

> No one nation or race of people has a monopoly on creative and inventive talent; but new ideas, like natural resources, don't amount to much unless and until something is done about them (Anonymous).

THE RESEARCH CAREERS FOR MINORITY SCHOLARS (RCMS) PROGRAM

Diversifying the SMET Workforce

NSF established the RCMS program to recruit minorities into SMET disciplines, to retain them through graduate school, and to encourage them to enter both academic and industrial SMET research institutions. RCMS offers interdisciplinary research opportunities to talented black, Hispanic, and Native American undergraduates. The specific objectives of the RCMS program are to stimulate the recruitment of underrepresented minority students above current levels and the following:

- to substantially improve the retention rates of undergraduate SMET students;
- to significantly increase the quantity and quality of minority students receiving undergraduate SMET degrees;
- to provide direct motivation and preparation for the entry of minority students into SMET graduate programs;
- to catalyze increased faculty involvement in issues related to SMET education.

RCMS was designed to catalyze sustainable change in practices that prevented full participation by individuals from minority groups underrepresented in SMET disciplines.

THE RCMS MENTORING EXPERIENCE

RCMS project directors and faculty mentors have served as advisors for more than 2,300 undergraduate SMET scholars. An impressive list of 57 research publications and 571 research presentations illustrates the effectiveness of involving undergraduates in faculty-directed research projects. These undergraduate scholars are doing research both on their campuses and on university campuses throughout the United States, Europe, Asia, and Africa. Their retention rate in SMET is more than 92 percent, with most RCMS graduates matriculating into prestigious graduate programs. The impact of the RCMS program will be felt for many years beyond the program's last award.

The RCMS program demonstrates how to effect fundamental change in the attitudes of administrators, faculty, and students regarding the involvement of undergraduates in SMET research projects and how to encourage the development of SMET student faculty mentoring relationships. The fundamental component of productive RCMS projects is the establishment of an environment that supports developing mentoring relationships between research faculty and undergraduates.

RCMS project directors report that student faculty mentoring relationships develop best when individual research faculty and students value their role in the relationships and commit their time to making the relationships work. Mentoring relationships require a large time commitment to be effective. One project director stated that:

Students, faculty, and administrators must recognize and value the time commitment needed to develop supportive student-mentor relationships (Bagayoko & Kelley, 1994).

The mentoring relationship provides many benefits for both mentors and students. Benefits for campuses that support developing mentoring relationships include increased retention rates for undergraduates, better SMET learning environments, better classroom experiences for all students, increased research productivity, and increased attractiveness of SMET-related career goals. University and college administrators can foster the development of mentoring relationships by helping the faculty and students with aspects of effective mentoring that unnecessarily consume faculty and student time.

- Administers should set clear policies that support faculty working with students and ensure that all SMET students receive some research experience before receiving a baccalaureate degree.
- Administrators should seek to intertwine mentoring activities with research experiences. In mentoring relationships the distinctions among teaching, research, and service are blurred. Just as faculty members are recognized for the quality of their teaching, research, and service, so should they be recognized for the quality of their work as mentors.

CREATING SUPPORTIVE RESEARCH STUDENT-MENTOR RELATIONSHIPS

Campus Administrative Support

Administrative support for faculty and students involved in mentoring relationships is critical for institutionalizing mentoring strategies. Mentoring should be considered a relationship among students, faculty, and administrators. All parties should view research mentoring as integral to undergraduate SMET student development. Administrators can strengthen campus mentoring by offering support to faculty and students, creating realistic expectations, and emphasizing results. Each of these points is discussed below.

Offering Support to Faculty

Tangible administrative support for active faculty mentors is essential to successful mentoring programs. Administrators should ensure that faculty have the resources necessary to implement mentoring relationships. Common examples of administrative support include providing release time for faculty mentors, additional research resources, travel support, special recognition programs, mentorship training, and assistance in recruiting prospective mentors.

Offering Support to Students

Successful mentoring programs include making undergraduate research experiences part of the undergraduate SMET curriculum, providing resources for student participation in SMET research conferences, and providing training in SMET research techniques and presentations.

Creating Realistic Expectations

Not all faculty make good mentors at all points in their careers. Successful mentoring programs do not force faculty members into mentoring relationships that they do not value; nor do they include students who are not committed participants. No one can mentor a student who does not want to be mentored.

Emphasizing Results

When a mentoring program achieves its short- or long-term goals, supportive institutions ensure that the accomplishment is adequately recognized. They develop methods for both recognizing and rewarding faculty and students who participate in effective mentoring relationships. Nothing is more damaging to the development of a strong mentoring relationship than an indifferent response to goal attainment.

Mentoring Office Support

Some aspects of mentoring are not discipline dependent. A campus office may be useful in the early development of mentoring activities. The mentoring office should provide the administrative infrastructure support and coordination to foster and monitor activities across the campus. In addition, the office should be responsible for developing a pool of experienced mentors, marketing the mentoring program, implementing an evaluation strategy, involving faculty and students in all key decision making, assisting discipline units in analyzing their capacity to participate in a mentoring program, and securing tangible commitments from both faculty and students before mentor/student relationships are formed.

Developing a Pool of Experienced Mentors

In successful mentoring programs, experienced mentors assist those who desire to become involved in mentoring relationships. They assist in student-faculty orientation activities by demonstrating how to build effective, supportive mentoring relationships. Experienced mentors also serve as role models for aspiring mentors.

Marketing the Mentoring Program

Mentoring offices can disseminate information to the faculty on the benefits of becoming involved in mentoring relationships. In addition, the mentoring office can serve as a source of information and training for both students and faculty interested in being participants in mentoring relationships.

Implementing an Evaluation Strategy

An acceptable evaluation strategy not only monitors the mentor/student relationship, but also reviews the policies of the administration to ensure that the program is receiving adequate attention from the administration.

Involving Faculty and Students in all Key Decision Making

This is particularly important for students. Being involved in decision making helps students feel more in control of their careers. The more students feel invested in their future, the more motivated they are to perform well.

Assisting Academic Discipline Units in Analyzing their Capacity to Participate in a Mentoring Program

Successful mentoring programs do not rush unprepared departments into implementing mentoring programs.

Securing Tangible Commitments from Both Faculty and Students Before Mentor/Student Relationships are Formed

Both faculty and students benefit from the mentoring experience, but if faculty or students are overcommitted to non-mentoring activities, the mentoring relationship suffers. Successful mentoring programs clearly define mentor and student roles and responsibilities.

Infrastructure Support

Successful mentoring programs pursue the development of institutional policies that support strengthening mentoring relationships. As part of the policy, successful mentoring programs should include the following.

Developing workshops. Workshops for prospective mentors and students should be offered on a regular basis. This is an effective way of recruiting capable mentors and students into the program and of providing publicity for the program.

Holding training seminars. Training seminars for mentors should cover such topics as minority issues, the concept of mentoring, the purpose of the program, and what can be expected throughout the mentor/student relationship. Implementing appropriate training can help create a successful mentoring program that benefits mentors, students, and their institutions.

Offering orientation sessions. Orientation sessions should be required for mentors and students once they are accepted into the program. The time should be used to elaborate on the mission of the program, to develop and establish clear and realistic goals that define the parameters of the mentor/student relationship, and to define roles and responsibilities.

A supportive mentoring infrastructure should also include organizing and supporting social activities. Evaluation strategies should be used to improve all aspects of the mentoring program.

RCMS INSTITUTIONAL MODELS

The following describe some of the outstanding examples of RCMS institutions that have discovered aspects of effective mentoring which works for their students. All of these schools have increased student retention within the sciences and engineering and have provided a strong, supportive environment to encourage undergraduates to become research scientists and engineers. Many of the graduates from these programs have completed vigorous undergraduate research work, published papers, made presentations at national conferences, and entered graduate school.

MENTORING IS NOT RESEARCH SUPERVISION

City College of the City University of New York

Program director: Neville A. Parker
Total number of students in program: 72
Total number of graduates: 32
Average GPA (1994): 3.25
Number of publications and presentations: 66

Since 1989 City College of the City University of New York's (CC of CUNY) RCMS program has cultivated a rich environment for students to practice their research presentation skills. Between 1992 and 1995, 46 students represented the program in the college-wide Scholars Day event, where they presented oral or poster presentations of their research. CC of CUNY's RCMS program has created a highly effective off-campus network involving industry and other research institutions.

Often, the real challenge to a student's performance is not academics, says Neville Parker, a program director of the RCMS program at CC of CUNY. This belief is the reason Parker advises RCMS mentors to consider social pressures as they advise students. Apparently, this philosophy of encouraging both professional and personal development works, because the RCMS program at CC of CUNY has expanded from an initial enrollment of 12 students during 1989-1990 to 54 students during 1993-1994.

Parker and his team of faculty mentors have carefully crafted their relationships with students. A mentor brings to the table the ability to work with students at their own pace; just having the opportunity to have a close relationship with someone in the sciences provides encouragement, he explained.

At CC of CUNY the key to the RCMS program's success is the view that mentoring is not the same as research supervision. There is a tendency to use the terms synonymously, but mentoring fully engages the student in what the mentor does and fully engages the mentor in what the student does, Parker says. Faculty mentors offer advice not only on lab work and research, but also on family and social issues.

The RCMS program actively encourages students to develop research skills early in their undergraduate careers, but it also strives to help students develop other skills required to become well-rounded students and scientists. For instance, RCMS mentors at CC of CUNY, through a joint effort with the university's speech department, encourage students to improve their communication skills, particularly their ability to make oral presentations.

Tracy DeJesus, a mechanical engineering major who entered the program in 1990, says that RCMS students at CC of CUNY have been very fortunate. She says her involvement with RCMS has made her a better student and a better person. Today, I am more motivated, more confident in what I can do, and more disciplined in dealing with problems and learning how to approach them, she says. DeJesus graduated in 1995.

When DeJesus entered the program she was unsure of her career aspirations. She says she decided on mechanical engineering simply because of a high school class she had taken. Her involvement with RCMS has motivated her to attend graduate school part time next fall while working in industry. "To me, the RCMS program is like a family unit; we all know each other." She says her mentor was an integral piece of this family unit who offered guidance and helped set her academic life on track.

Parker says that this type of encouragement is what keeps students in the RCMS program and, more importantly, in the science and engineering pipeline. Of the 32 graduates, 22 have been, or are going, to graduate school. Three are currently pursuing their doctoral degrees.

MENTORS FILL RESOURCE GAPS

Grambling State University

Program director: A.N. Murty
Total number of students in program: 39
Total number of graduates: 9
Average GPA (1989-1994): 3.6
Number of publications and presentations: 20

Grambling State University's (GSU) RCMS program has been crucial to the academic progress achieved by the school's physics and chemistry departments. With the program's help, both departments have increased enrollment and retention, and of the nine students who have graduated from the program, seven have enrolled in graduate school. In addition, GSU has established excellent undergraduate research facilities that have made it possible for students to present more than 60 research papers at local, regional, and national conferences. Physics major Ursula Donatto received first place at the NSF Diversity Conference.

"Diversity is an absolutely necessary and desirable component for any type of revolution, be it scientific or otherwise," says A.N. Murty, program director of the RCMS program at Grambling State University. A revolution is precisely what has occurred at GSU since the implementation of the RCMS program in 1989. Before the program's inception GSU graduated an average of only four students per year in physics and chemistry combined. Today, that average has more than doubled to 10 per year. In addition, the total enrollment in both of these majors has nearly doubled since 1989.

Murty believes that the strength of the program lies with its faculty members. Their involvement and enthusiasm to motivate and sustain the zeal of the students plays a vital role. Mentors spend substantial individual time with students and attempt to be more than simply their research advisors. Mentors don't only go through the academics, they must also address other concerns of the students and get involved in their personal lives if need be, Murty says.

Tanya Lewis, a senior who graduated in December 1995, says that her relationship with her mentor has helped her through both personal and academic challenges. Lewis, who began her involvement with RCMS during her freshman year, took a full semester off from school for personal reasons. During that time, she said her relationship with her mentor and the love of research she developed through RCMS helped her keep her resolve to return to school. "I remember sitting in my house and thinking about what it used to be like to get up and go to school everyday and do research. I realized then that I had a gift, and I missed it. I knew that I wanted to spend my time doing research." The next semester she returned to school with a new determination and made plans to attend graduate school beginning in the fall of 1996.

The RCMS program at GSU emphasizes such personal involvement between mentors and students. Murty believes that the minority students that this program targets are in particular need of mentors and role models. "Of course, everyone needs mentors, not just minorities, but minorities are so often coming from environments that do not provide the necessary resources, and the mentor will fill that gap."

In addition to the heavy emphasis on the mentoring aspect of the RCMS program, students at GSU are also provided with ample research opportunities in a variety of laboratories. Students are encouraged to work on independent endeavors and to support their research advisors' projects. The overall goal is to encourage students to develop the desire and discipline needed to make research a personal goal as well as an academic one. Lewis says that her involvement with the program helped her discover a love of research that she did not have before her college experience. Apparently, this attitude is prevalent because more than 50 percent of RCMS program participants go on to graduate school. This all adds up to the true measure of success for RCMS.

MENTOR BY EXAMPLE

Howard University

Program director: Winston A. Anderson
Total number of students in program: 69
Total number of graduates: 23
Average GPA (1991-1994): 3.23
Number of publications and presentations: 4

Howard University's RCMS program, which began in 1989, trains undergraduates in research disciplines, such as engineering and the physical and natural sciences. Trainees apprentice at several Centers of Excellence, including Material Science Research Center of Excellence; Computational Sciences and Engineering Research Center; Center for the Study of Terrestrial and Extraterrestrial Atmospheres; Laser Chemistry Laboratory; and Particle Detector Research Center. Through an international exchange program, RCMS students have access to seven leading European and Asian laboratories and 19 research universities sponsored by the Leadership Alliance. The university also sponsors the Distinguished Scientists Seminar series which enhances the university's intellectual environment. High school teachers attending the university's summer continuing education program have developed "Molecules in our Lives," a manual designed to improve precollege chemistry education.

The RCMS program at Howard University supports 20 to 30 undergraduates each year and maintains a retention rate of more than 95 percent. Most of the students who have participated in the program have gone on to graduate school in the science, engineering, and mathematics fields. One of the most important aspects of the university's success is mentoring, says Winston Anderson, program director. The mentor is the first line of excellence and vision that the student observes. Mentors offer essential, one-to-one, direct counseling and career advice. Indeed, Anderson encourages faculty mentors to spend substantial time with their students because, without mentoring, many minority students would not be exposed to the best opportunities the school has to offer.

Faculty mentors say they gain both personal and professional fulfillment from their involvement in the RCMS program. "Working with students gives you the energy to keep going," says Folahan Ayorinde, an associate professor of chemistry. Ayorinde believes that the emphasis on mentoring as a relationship between peers is what makes the program at Howard so successful. "You can't just tell [students] what needs to be done, you have to mentor by example and realize that you learn as much from them as they learn from you."

Joseph Spence, a fourth-year electrical engineering student, said that he began participating in RCMS strictly because of the research opportunities it offered. Spence, who has been working with the RCMS program for two years, says that the experience he has gained has been invaluable. "I have found the opportunity to be more creative and exercise muscles that I have not used in the classroom. The greatest accomplishment so far has been that I now feel comfortable expressing my opinions and ideas in a circle where everyone else has their Ph.D." Spence says mentoring is vital to the success of the program. "I couldn't see a program like this running without mentors. My relationship with my mentor is so personal that talking to him is almost like talking to myself."

The program at Howard has been immensely successful in building relationships like the one between Spence and his mentor. The combination of opportunity and dedication has facilitated these relationships that will remain important for years to come.

MENTORING REQUIRES AMPLE TIME

Jackson State University

Program director: Abdul K.A. Mohamed
Total number of students in program: 47
Total number of graduates: 33
Average GPA (1993-1994): 3.42
Number of publications and presentations: 17
*Includes only students who were RCMS scholars in 1993-1994.

Jackson State University is one of the few historically minority institutions in the nation with an environmental and marine sciences program. The university also has outstanding meteorology and mathematics departments. RCMS students take special courses that consist of supervised research projects, laboratory experiments, and presentations of their findings. Students travel to at least one conference or meeting related to their research experiences. The retention rate has been 98 percent.

RCMS is bigger than Jackson State, says 1994 program graduate, Christopher Hall. Hall, who is currently enrolled in graduate school studying molecular biology, means that the program at Jackson State is involved with students both on and off campus. RCMS at Jackson State requires all of its upperclass participants to tutor campus students five hours per week. In addition, RCMS sophomores take part in a program that provides tutorial services for local high school students. All of these efforts have not only helped countless students, but have also boosted the program's recruiting efforts. Participation in the RCMS program at Jackson State has more than doubled since its founding in 1991.

Program director, Abdul K.A. Mohamed, credits much of the program's success to the linkages established between RCMS and other programs designed to encourage minority participation. The financial support that supplements the NSF grant allows students to dedicate much of their time to research, rather than part-time jobs. Mohamed says that one of the most satisfying aspects of the program is having the ability to help students who might otherwise struggle through such rigorous work.

Faculty mentor Dan Course agrees. "There are some students that you know can succeed, but a little urging would help them. They need to be nudged onto higher ground to flourish," he says. Another faculty member, Arthur Jones, who has not served as a mentor but has taught many RCMS students in the classroom, says that faculty members throughout the university enjoy working with RCMS students. "We're excited about working with RCMS students; in them we have the opportunity to see the first real spark of success at the undergraduate level," he says.

Students who participate in RCMS at Jackson State have the opportunity to perform research at the undergraduate level during the school year and over the summer. Almost every RCMS student participates in summer research projects across the country. Involvement in this research is considered one of the most valuable aspects of the program by many of the students. When Hall entered the program he was unsure whether he wanted to move on to medical school or to a doctoral program. Hall attributes his decision to pursue graduate studies in molecular biology to his involvement with RCMS.

The mentoring component of the program is also vitally important to its participants. Active faculty participation has enabled the program to keep its faculty mentor ratio at 2:1, providing each student with a mentor who has ample time to devote to his or her development. The strong relationships between mentors and their students have also boosted the program's retention rate, which stands at 98 percent. The combination of the efforts of the faculty, the program director, and the students at Jackson State has built an outstanding program.

MENTORS SERVE AS GUIDES, SHOWING STUDENTS HOW TO BALANCE A CAREER AND A SOCIAL LIFE

Navajo Community College

Program director: Mark C. Bauer
Total number of students in program: 31
Total number of graduates: 17
Average GPA (1992-1994): 2.72
Number of publications and presentations: 8

Since 1992 Navajo Community College's (NCC) RCMS program has assisted Navajo students in achieving a successful transition from reservation high schools to the university. The program has increased the number of science majors at the college by about 20 percent. NCC has developed collaborative academic ties with several science and technology institutions. In addition, NCC has placed all students in summer research experiences.

Eldon Blueyes is spending his summer vacation performing research at Ahmadubello University in Zaria, Nigeria. This is Blueyes's first trip outside the United States. He is looking forward to learning about biology in another country, learning about the Zarian culture, and forming professional contacts for the future. The research in Nigeria is Blueyes's summer placement, a major focus of the Navajo Scholars Program at NCC.

Each summer NCC places students in university and industrial research programs at institutions such as New Mexico State University, the University of New Mexico, and the University of California's Lawrence Livermore National Laboratory. The placements offer many students their first opportunity to leave the reservation, and serve as professional experiences that complement academic studies. During the past three and a half years these placements have been extremely successful, says program director Mark Bauer. NCC has not only placed every one of the 20 students in the RCMS program, but has also placed 20 students outside the program. The labs have been so hungry (for students) that we have more than doubled our original placement goal, Bauer says.

Professional experience is not the only benefit students receive from summer research opportunities. With each placement students are assigned summer mentors to help guide them. These mentors typically spend more than just a summer with their students. Blueyes is still in contact with Greg Lennon, a mentor from his first placement at the Lawrence Livermore National

Laboratory. Blueyes credits Lennon with showing him that a balance between a career and a social life can be achieved.

The Navajo Scholars Program provides more mentoring support than most other mentoring programs; it assigns two mentors to each student. Blueyes's second and main mentor is Vama Robson, a biochemist who takes a personal interest in all her students at NCC. Robson teaches the Orientation to Research class that introduces students to statistics and research and helps them with applications for summer research positions. She teaches students specific skills about how to use research tools and to present papers. For example, Robson videotapes students as they present research findings to the class to help them improve their presentation. In addition, Robson invites guest researchers and encourages her students to question them about specific research projects and tools.

The Navajo Scholars Program is one of the most successful RCMS projects, with a retention rate of 85 percent and 70 applicants for the 32 positions. Its principal objective is to assist Navajo students with their transition from high school on the Navajo reservation to four-year universities. With this in mind, the program seeks to provide the necessary financial support and to strengthen students' communication, mathematics, and science skills, and their confidence and motivation. "Many of our students haven't been around people with advanced degrees. They come to the college to obtain a degree, but they don't really know what it takes. They haven't been exposed to scientists," Bauer says.

MENTOR PROGRAM DIRECTORS MUST GET TO KNOW MENTORS AND STUDENTS

Oklahoma State University

Program director: George Dixon
Total number of students in program: 36
Total number of graduates: 7
Average GPA (1990-1994): 3.6
Number of publications and presentations: 38

Oklahoma State University (OSU) has increased the enrollment of students in the RCMS program since its beginning four years ago. Students have participated in a variety of original research efforts, including laser research, that continues into graduate school and future employment. The program has increased enrollment of minorities not only in science and technology fields, but also at the university.

In the past decade OSU has awarded more science and engineering doctorates to Native American students than any other university in the United States. The RCMS program has helped OSU strengthen the university's dedication to undergraduate research and science, engineering, and mathematics education and has encouraged minority participation. Since its implementation in 1990 enrollment in the RCMS program has more than doubled, the retention has remained high, and participants have maintained a collective grade point average above 3.6.

Program director, George Dixon, says that encouraging minorities, especially Native Americans, to study science is vitally important. For example, environmental science becomes particularly important to Native American students when federal and tribal governments negotiate issues about tribal land, Dixon says. American Indians want to be able to represent their people in these decisions. The RCMS program at OSU is helping minority students meet many such personal goals.

Much of the program's success can be attributed to the early research opportunities it offers students. "Research is habit forming. It's addictive. By getting these students involved in research early in their careers, we hope to get them hooked. So far, all but two of the RCMS participants at OSU have gone on to pursue advanced degrees," Dixon says.

Michael Boone, a 24-year-old physics major, says his studies have been greatly enhanced by the RCMS program. Boone says the greatest advantage of the program is the chance to work side by side with other minorities that have the same skills as he has. Before he became involved with RCMS Boone says he planned to get his bachelor's degree and enter the workforce; now he plans to pursue a master's degree in physics.

At the heart of this type of success is a careful mentoring relationship between students and faculty members. Dixon spends much of his time getting to know prospective mentors to ensure their dedication to the program. "You need to know your mentors and feel confident that they will actually spend time with the students," he advises. "This is a vital part of the RCMS program, and all of the time and hard work is well rewarded. Because of our program and others like it, we have seen an enormous growth in minority student enrollment in the sciences."

MENTORS SHOW OVERALL CONCERN FOR STUDENT DEVELOPMENT

Southern University and A. & M. College, Timbuktu Academy

Program director: Diola Bagayoko
Total number of students in program: 40
Total number of graduates (as of 5/31/95): 14
Total number of graduates (in 1995-1996): 8
Average GPA (1993-1994): 3.30
Number of publications and presentations (1994-1995): 19

Southern University's Timbuktu Academy has developed into a successful tool for encouraging and recruiting minorities into the science, mathematics, and technology fields. The school encourages students to attend and participate in conferences and research projects.

Since Southern University established its Timbuktu Academy in 1990 the RCMS program has compiled impressive statistics. For example, more than 80 percent of RCMS students go to graduate school, physics enrollment has increased from 20 to 70 students over five years, and more than 90 percent of the physics students are engaged in summer research at national laboratories. Integral to the program's prosperity is its philosophy that students and mentors alike are responsible for the success of the program. Program director, Diola Bagayoko, says that the formula for successful mentoring involves more than financial support. Mentoring means providing other types of support from a variety of sources so that the student can have enough time to devote to conferences, research, and so on. If students do not have support, then they will not have time to invest in the program.

These other types of support are evident in Southern's vigorous monitoring system, a regimen that includes a bimonthly review of transcripts, the updating of student curriculum vitae every semester, guidance on course selection and sequence, and personal counseling. Such vigilance is important in order to ensure that students are progressing in the program. It also allows mentors to encourage students who are doing well. "If I do not monitor, how am I going to be able to tell that [students] have done a great job?" Bagayoko says, "Although the program has a great deal of structure, the relationship between the mentor and the student is left to develop both on a personal and professional level." Wilson Sheppard, one of Bagayoko's research assistants since 1991, says Bagayoko showed overall concern for his development. "He made sure I was not only performing well academically but also in my private life. He commended me when I did well, and he was there to discipline me when I needed it."

A basic tenet of Southern's program is that participation in research should begin in the freshman year. "Students must do hands-on work...they should work in the lab rather than work at fast food restaurants," says Rambabu Bobba, associate professor in the physics department. Another tenet of the program is that all students should attend professional conferences. Conferences are an important arena for learning and making professional contacts, Bobba says.

Sheppard, who graduated from the program in 1994 and is pursuing his master's degree in physics, found that conference participation was valuable in his professional development. "I had the opportunity to meet my peers and explore my knowledge. Knowing where you stand in the scientific community and presenting technical papers are important to your development as a scientist."

The objectives, paradigm, programs, activities, and selected results of the Timbuktu Academy are described in articles by Dagayoko and Kelley (1994) and Moore and Bagayoko (1994).

MENTORING IS A SORT OF APPRENTICESHIP, A POWERFUL FORM OF LEARNING

State University of New York at Stony Brook

Program directors: David L. Ferguson and Alan Tucker
Total number of students in program: 91
Total number of graduates: 2
Average GPA (1993-1994): 3.21
Number of publications and presentations: 2

State University of New York (SUNY) at Stony Brook has had great success with the RCMS program and its participating students. Prior to instituting the RCMS program in 1992 only about 15 minority students in science and engineering majors at Stony Brook had grade point averages of 3.0 or higher. Currently, there are 100 such students.

"I hadn't considered graduate school as an option... [my mentor] is probably one of the main reasons that I'm going to graduate school," exclaims mathematics scholar Cesar Mena, a 1995 graduate of SUNY at Stony Brook and a participant in the RCMS mentoring program. Mena currently plans to continue his study of mathematics in graduate school at Stony Brook. He says that the RCMS program not only has had obvious effects on his academic career, but also has helped him develop social relationships with other minorities his age. This type of comprehensive support and guidance is the goal of Stony Brook's RCMS program.

The role of a mentor is equivalent to the role of an athletic coach, says David Ferguson, a professor of technology and sociology, and applied mathematics and statistics at Stony Brook. Both use an experienced person to guide student development. Mentoring is a sort of apprenticeship, a very powerful form of learning, Ferguson says. Mentoring offers students access to the wide variety of human and material resources available on an academic campus.

Ferguson co-directs the RCMS program with Alan Tucker, a distinguished teaching professor in applied mathematics and statistics. Under their leadership the program has enjoyed tremendous success since its inception in 1992, beginning with 22 hard-working students and currently serving 120 students in mathematics, engineering, physics, and chemistry. Program coordinator, Michele McTernan, attributes part of this success to Stony Brook's extensive network of support for minority students. "We can offer students opportunities and resources that may not be available elsewhere," she says.

The program's rapid growth is also the result of a great deal of hard work and effort. For instance, Ferguson says that a successful mentoring program encourages strong faculty input in the planning and operation of the mentoring program. "It is critical... to publicize the success of the program to all science, mathematics, and engineering departments. When the faculty see the real impact that the program has, more of them will want to participate as mentors."

Stony Brook no longer has to recruit mentors into the program; instead, faculty members volunteer to be mentors, calling the program office to request students for research and training, Ferguson says. "It's a great situation, and it's good for both faculty and students."

A driving force behind the program is its emphasis on active learning and strong student participation in the education process. Ferguson directs calculus workshops and says he rarely sees students sitting passively, waiting to be filled with knowledge. Tucker agrees. He says that the academic activities engage students in cooperative learning and gives them a glimpse of the real world which awaits after graduation. The academic support activities permit undergraduate students to work together, with the support of graduate students and university faculty, just the way they would work as part of a team in industry.

(A MENTOR IS A) PERSON YOU CAN TURN TO WHEN YOU WANT TO MAKE THE CRITICAL DECISIONS

Texas A&M University

Program director: Karan L. Watson
Total number of students in program: 173
Total number of graduates: 70
Average GPA (1993-1994): 3.36
Number of publications and presentations: 59

Texas A&M University is making remarkable progress in the RCMS program. The enrollment of minorities has increased, along with the number of minorities entering the science, mathematics, and technology fields. Half of the students in the program have pursued their graduate degrees in science and technology fields. Those who have not gone to graduate school have taken positions at various research companies and government agencies. One student enrolled in the navy as a nuclear power officer, and two are pursuing teaching degrees. Oscar Serna won the NASA Graduate Student Researcher Program scholarship.

Karan Watson, director of the RCMS program at Texas A&M, knows firsthand what being a minority student in the sciences is like. "As a woman in engineering, I understand the need for programs like RCMS. It was difficult being a woman student because it was clear I wasn't getting the same information as male students were through their personal relationships with faculty." The RCMS program at Texas A&M attempts to build such personal relationships between minority students and faculty so that "there is someone there [for the students] with more experience to suggest options," Watson says.

The program at Texas A&M has more than tripled in size since its inception in 1989. A key factor in this substantial growth rate has been the emphasis on the mentor/student relationship. Participants have the opportunity to select their own mentors so that each student is matched with a faculty member with similar interests. The program strongly encourages students to meet and interview several research faculty members before selecting a mentor.

Enos Inniss, a 1995 civil engineering graduate, says that his mentoring experience was a maturing one. "I see a mentor as the first person you can turn to when you want to make the critical decisions that your life will be based on." Inniss, who plans to pursue his doctorate in environmental engineering at Notre Dame, became an RCMS participant to help learn what graduate school would be like. He says his research experience was invaluable for many reasons, but particularly because it helped him narrow his field of study.

Inniss says that another aspect of RCMS that contributed to his overall satisfaction with the program was the required outreach project. Texas A&M requires junior and senior participants to serve as positive role models through their involvement in two public school outreach experiences. These experiences are designed to encourage younger students to choose engineering and science as careers. Inniss found the outreach program so fulfilling that he always did more than the required to. "I enjoyed making younger students aware of what was available in engineering," he says.

Watson says that the combination of these aspects have made the RCMS program at Texas A&M successful for the entire university. There is a large number of minority students who have visions of going on to graduate school and doing research. And that group has expanded beyond the RCMS program, she says.

MENTORING IS A TWO-WAY STREET

University of California, San Diego

Program director: David McDonald
Total number of students in program: 133
Total number of graduates: 49
Average GPA (1990-1995): 3.14
Number of publications and presentations: 2

The University of California (UC), San Diego RCMS program promotes science education among minorities of all ages and it has strengthened ongoing efforts to improve science education in San Diego high schools. As students move on to college the program sponsors minority undergraduates in scientific research and encourages them to pursue graduate study in oceanography and other scientific fields. In addition, the Scripps Undergraduate Research Fellows (SURF) program offers a 10-week summer research experience for more than 20 minority undergraduates each year.

Every summer the coast of Southern California near San Diego faces an onslaught of eager undergraduates striving to learn more about oceanography and other science fields. These students spend 10 weeks working closely with a research supervisor and experiencing what one student calls life as a graduate student. Students divide their time between individual research projects, weekly lectures, GRE preparatory classes, and graduate school application workshops. The SURF program strives to encourage undergraduates to pursue advanced studies in the sciences, and RCMS provides funds for an average of 20 such future scientists each year.

The program at UC, San Diego, is unusual because it involves precollege, undergraduate, and graduate education. At the precollege level UC, San Diego, actively recruits minority students through local connections and the already established Scripps Aquarium, which aims to foster youth interest in marine science. The graduate program helps support up to three minority students each year in their study of marine science. UC, San Diego, which is one of the few institutions in the country that offers this program at the graduate level, considers the RCMS support of graduate students crucial to the success of the program's goal to increase minority participation.

At the undergraduate level the program encourages minority students to pursue their interest in marine science through the summer SURF program, a faculty mentor program, and participation in undergraduate research conferences. Kathleen Avina, a SURF student in 1991, says she values those experiences. "I left San Diego knowing more about my own abilities to participate in science and be part of an academic community."

David McDonald, director of Graduate Student Affirmative Action Programs and the RCMS program, strongly believes in the mission of RCMS. His conviction carries over to the strength of the program. To McDonald, diversity is important in science, engineering, and mathematics research because science is about diversity. We're talking about avoiding replication of flawed principles. Diversity encourages academic and research integrity.

Mentoring, McDonald says, remains one of the most crucial aspects of the program. It is one that he believes to be multifaceted in its benefits. Mentoring is a two-way street, it allows the student to have a link with lofty research goals, and it allows the mentor the opportunity to connect back with the students and learn from a new generation of young scholars.

The efforts of the RCMS program have been well rewarded. More than 80 percent of former participants have been admitted to graduate programs across the country, and minority applications to the UC, San Diego, Scripps

Institution of Oceanography have more than tripled since the program began. At the same time, the number of enrolled minority students has more than doubled.

MENTORING PROVIDES STUDENTS WITH A VARIETY OF ROLE MODELS AND MUCH-NEEDED SUPPORT

University of Maryland Baltimore County

Program director: Freeman A. Hrabowski III
Total number of students in program: 180
Total number of graduates: 47
Average GPA (1992-1994): 3.4
Number of publications and presentations: 40

The University of Maryland Baltimore County (UMBC) Meyerhoff Scholars Program, instituted in 1988 and partially funded by RCMS, is highly successful. The program enrolls 180 students and has a retention rate of nearly 100 percent. One of the program's key components is the opportunity students have to engage in summer research experiences at some of the nation's leading universities and top scientific, computing, and technological companies. Forty-three of the program's graduates are currently enrolled in Ph.D., M.D., or M.D./Ph.D. programs.

In May 1993 the six young black men who crossed the stage at graduation at UMBC had earned more than a bachelor's degree. They also had earned a feeling of personal accomplishment by defying the odds and had developed the ambition to have successful careers in scientific research. These young men were the first graduates of the Meyerhoff Scholars Program, which strives to increase the number of black students earning doctoral degrees in science, engineering, and mathematics. The Meyerhoff program is partially funded by RCMS.

Mentoring is an important component of the Meyerhoff program. Freeman A. Hrabowski, UMBC's president and co-founder of the Meyerhoff program, says, students need to know that others like themselves have succeeded in science and engineering mentoring provides students with a variety of role models and much-needed support.

In addition to mentoring, the Meyerhoff program's other key components make it one of the most comprehensive and distinctive programs of its kind in the nation. These components include (1) recruitment of top students to UMBC which received more than 700 applications for the 47 scholarships awarded to new freshmen enrolling in the fall of 1995; (2) financial support for the scholars, including full tuition and fees, room and board, and a $1,000 book allowance each year; (3) a six-week, residential, summer bridge program for all new students before their freshman year that gives the students a solid foundation for success in the fall by orienting them to university-level work and by encouraging them to develop special bonds with each other; (4) emphasis on the value of study groups and tutoring; (5) encouragement of a strong sense of community and of programmatic values among the scholars; (6) summer research internships in university and corporate labs leading to research presentations and publications; (7) the faculty and administration's active involvement and strong support; and (8) family involvement, especially through the Meyerhoff Program Parents Association.

Since its inception in 1988 the Meyerhoff Scholars Program has been highly successful, graduating 47 students. Forty-three are currently enrolled in Ph.D., M.D., or M.D./Ph.D. programs at nationally recognized universities (and other graduates intend to enroll). In the 1995-1996 academic year UMBC projects the highest number of Meyerhoff graduates to date. Dr. Hrabowski says that the Meyerhoff program's success in recruiting and graduating talented blacks in science, engineering, and mathematics (SEM) is vital. "Diversity in SEM research is important to the economic well-being of our nation. Because an increasing proportion of the workforce will be minorities, we need to make sure the workforce is scientifically and technologically prepared. With funding from the National Science Foundation's RCMS Program, UMBC is addressing the critical shortage of minority students who enter doctoral programs and have research careers in the sciences."

WHEN IT (MENTORING) WORKS, IT'S SOMETHING THAT LASTS FOREVER

University of Puerto Rico

Program director: Brad R. Weiner
Total number of students in program: 45
Total number of graduates: 21
Average GPA (1991-1994): 3.53
Number of publications and presentations: 22

The University of Puerto Rico (UPR) has had enormous success with the RCMS program. During 1993-1994, 10 of 11 students who graduated enrolled in or sought admission to graduate programs in science. No students dropped out, and the minimum GPA was 2.68. Two Puerto Rico-RCMS meetings were organized during 1993-1994 with the participation of faculty members and RCMS students. During these meetings students presented their research results and attended a conference given by young local scientists.

When Felipe Acosta joined the RCMS program at UPR in January 1992 he knew he wanted to go to graduate school someday, but he was unsure if he was ready to commit himself to the extensive time involved. His involvement with the RCMS program has encouraged him to accept the challenge of a doctoral program in civil engineering. Acosta is currently in his third year of graduate school at the Georgia Institute of Technology. Many UPR RCMS students are following along a similar path; 96 percent of all of the program's graduates are pursuing graduate studies in scientific fields.

Mentoring, says program director Brad Weiner, is the key to UPR's success. Weiner recruited faculty mentors who were involved in intensive research programs and willing to take the time to form meaningful relationships with undergraduates. Weiner said that he feels mentoring is essential in the sciences. "Science is not something you learn in the classroom, it's a way of life, and most scientists who are involved in it understand that explicitly. It takes a lot of time to teach students that, and the mentoring relationship helps this process. When it works, it's something that lasts forever."

Stephan Gomez, who began a graduate program in chemistry in the fall of 1995 at Cornell University, says that her mentoring relationship has been integral to her success. "I could go to [my mentor] for advice or if I needed information and wanted to see where I stood." This type of interaction is exactly what the program strives for.

The program also strives to prepare students to perform research projects and presentations. Every RCMS participant is required to present research to faculty and peers on a regular basis. Acosta says he found this type of practice and training to be very useful. "I liked the challenge and the fact that we all had to share what we were doing and give our ideas." Some UPR students are even given the opportunity to present their findings to a broader audience by attending national and regional conferences.

Gomez also found the opportunity to attend these national and regional conferences quite helpful. She says that her attendance at such conferences heightened her interests in interdisciplinary research. They gave me the opportunity to meet with my peers and share different kinds of knowledge, she says. That interaction with her peers was the most positive RCMS experience for Gomez. "When I first started the program, I wasn't sure what field I wanted to pursue my Ph.D. in. But sharing with people in other majors and seeing how what they were doing related to what I was doing was very helpful."

REFERENCES

Adams, S.G., & Adams, H.G. (1993). *Techniques for effective undergraduate mentoring.* College Park, MD: National Consortium for Graduate Degrees for Minorities in Engineering and Science, Inc.

Bagayoko, D., & Kelley, E. (1994). The dynamics of student retention. *Education, 115* (1) 31-39.

Clinton, W.J., & Gore, A., Jr. (1994, August). *Science in the national interest.* Washington, DC: Executive Office of the President, Office of Science and Technology Policy.

Canto, M.E., James, D.P., & Brescia, W. (1994). *RCMS: Mentoring guide for community colleges.*

Fass, P. [1989]. Outside in: Minorities and the transformation of American education. New York: Oxford University Press.

Gibbons, J.H.. (1995). Speech to American Association for the Advancement of Science forum on Unity in Diversity. Washington, DC, June 26.

Jones, M.M. et al. (1987). *Chemistry and society* (5th ed.). Philadelphia, PA: Saunders College.

Moore, W.E., & Bagayoko, D. (1994). A paradigm of education: The model of the Timbuktu Academy. *Education, 115* (1), 11-18.

National Science Board. (1983). *Educating Americans for the 21st century.* Washington, DC: NSF.

REFINING AND EXPANDING THE ROLE OF PROFESSIONAL ASSOCIATIONS TO INCREASE THE POOL OF FACULTY RESEARCHERS OF COLOR THROUGH MENTORING

Stafford L. Hood and Jennifer Boyce

INTRODUCTION

There continues to be a shortage of educational researchers of color during a time when the needs for their presence and involvement in seeking answers to important research questions which will in many ways determine the futures of children of color in American society. Their presence and involvement in this effort are not luxuries, it is a necessity to our collective future as we enter the twenty-first century. Frierson (1990) has persuasively argued that more educational researchers of color are needed because they are an invaluable and largely untapped resource in the effort to improve the academic success of students of color in our schools. Unfortunately, as the number of children

Diversity in Higher Education, Volume 1, pages 141-159.
Copyright © 1997 by JAI Press Inc.
All rights of reproduction in any form reserved.
ISBN: 0-7623-0086-8

of color increases in our educational system, the number of new educational researchers of color has shown minimal growth and has probably declined since the 1970s if the rate of doctorate degrees awarded to people of color is used as a barometer (Frierson, 1990; Hood & Freeman, 1995).

The credentialing of educational researchers of color through doctoral programs at research universities is one part of the equation to increase their numbers, however, their ability to contribute to the research base is also dependent on the successful development of a career in educational research after the degree is received. The acquisition of the doctoral degree is insufficient to have a successful career in educational research unless the person is guided or advocated by someone who is already respected as an educational researcher—a mentor. The presence of a mentor during doctoral studies and in career development is important for anyone pursuing a career in educational research. The mentoring process, however, may be viewed as even more critical in the training and career advancement of future educational researchers of color.

Mentoring of educational researchers has been typically discussed within the context of doctoral study and junior faculty at research universities. Blackwell (1987), Epps (1989), Frierson (1990), and Moore (1988), to name only a few, ignited our continuing discussion on mentoring in the university setting. Still, it is important to recognize that the responsibility for mentoring doctoral students and junior faculty has typically been placed on faculty within the university where the individual is enrolled as a student or employed as a junior faculty member. In view of the shortage of faculty of color at our major research universities (Hood & Freeman 1995), doctoral students and junior faculty of color must seek mentoring from white researchers (probably male), researchers of color from other universities, and through professional associations. The mentoring of educational researchers within the setting of professional associations (that emphasize educational research) has not yet been seriously discussed in the literature. Even though such mentoring has been mentioned both formally and informally, structured efforts to mentor future educational researchers of color has been mostly unsystematic and inconsistent.

This chapter is intended to begin a more serious discussion which will hopefully lead to considerations for the development and implementation of more structured efforts to mentor future and new educational researchers within professional education associations. Although not often thought of in such a fashion, professional educational associations can play a more integral role in the career development and advancement of educational researchers of color.

The chapter has been organized into four major sections. The first section explores the true meaning of mentoring based on the origins of the term and evolving definitions of the participants (mentors and protégés), and nature of the mentoring relationship. The second section reiterates the important issue of the continued shortage of faculty of color at our universities (particularly research-oriented universities) and how this has negative consequences for the mentoring of people of color seeking careers as university faculty/researchers. The chapter then turns to a discussion as to how professional associations can maximize their potential as vehicles to increase the number of faculty/researchers of color with mentoring being an intricate component of this effort. The chapter finally explores possible strategies that professional associations may consider in an effort to refine what they may already be doing to facilitate mentoring within the association. While the American Educational Research Association will be a frequently used as an example throughout this chapter, this association is used only to frame the discussion for broader consideration across professional associations and particularly those that emphasize research.

UNDERSTANDING THE TRUE MEANING OF THE MENTORING RELATIONSHIP

Mentoring has a long history with the origins of the word "mentor" being translated from Greek mythology. Mentor was the tutor to Odysseus's son (Telemachus) and became his "sage guardian" when appointed by Odysseus upon his departure for the Trojan War (*Funk and Wagnalls' Dictionary*, p. 796). The dictionary further defines the mentor as a "...wise and trusted teacher, guide, and friend; an elderly monitor or adviser" (p. 796). Both statements confirm the notion of "trust" based on the mentor's dual role as teacher and friend. Yet it is further suggested that the mentor should also possess wisdom that is assumed to come with experiences acquired as a result of longevity. While there is a reasonable relationship between age and wisdom, age is no guarantee of wisdom. The important point to be made is that mentoring is intended to occur within the context of one person educating another. But there is also the expectation that the mentor "watch over" the recipient of this education. Consequently, education and protection of the protégé can be viewed as the basic essence of the mentor's responsibility. However, we have seen the role to both educate and protect the protégé expanded over time due to its application in numerous settings outside traditional school/university environment. Mentoring has clearly evolved far beyond the confines of a relationship where the mentor/teacher's sole objective is to transmit academic knowledge that is to be learned by the student.

DEFINING THE ROLE OF A MENTOR

There tends to be some consistency in the basic definitions of the mentor and what is considered to be the roles and expectations of such a person. Leon (1993) defines the mentor as a person of superior status and achievement who guides the personal and professional development of another person (protégé).

In this relationship the mentor may perform all or some of the following: teach, advise and counsel, assist and support, motivate and encourage, challenge, sponsor and facilitate a career, offer constructive criticism, and protect the protégé (Leon, 1993; Spray, 1994). Harris-Schenz (1990) eloquently wrote that a "mentor is a guide who leads a traveler on a pilgrimage of discovery" (p. 19). Atkinson, Casas, and Neville (1994) similarly define the mentor as:

> ...a trusted and experienced supervisor or advisor who by mutual consent takes an active interest in the development and education of a younger or less experienced individual.... A mentor proactively seeks to enhance the development and education of a protégé... (p. 39).

The role of a mentor implies a close, long-term, one-on-one relationship. But what is actually required of a mentor? A "good" mentor starts by showing an avid, genuine interest in the protégé. He or she should take time to learn about the protégé's personal background and interests but place the primary emphasis on the protégé's academic and/or occupational progress (UCLA Graduate School of Library and Information Science, 1991). Below is a list of some of the important duties of a mentor that can be generalized across settings (Adams, 1992; UCLA Graduate School of Library and Information Science, 1991, p. 16; Valadez & Duran, 1991).

1. Teaching and guiding (protégé must make own decisions);
2. Advising and counseling (on formal and informal matters of the profession).
3. Sponsoring and protecting (being an advocate; promoting protégé in profession);
4. Role modeling (setting a good example);
5. Validating (acknowledging achievements and giving sound, constructive, and critical feedback);
6. Challenge (exposure in new skills and projects);
7. Motivating (helping protégé to meet challenges);
8. Encouraging (providing emotional support and helping combat stressful periods);
9. Communicating (being open, spending time, and listening);
10. Invisibility (being subtle and not accepting credit); and
11. Identify professional contacts and networking opportunities.

The above is probably best summarized as an investment of time and resources in the development and promotion of a viable academic and professional candidate (Adams, 1992). There should be considerable latitude in the roles and responsibilities of the mentor and protégé but the expectations of each should be articulated and decided prior to entering the mentoring relationship (Phillips-Jones, 1983; UCLA Graduate School of Library and

Information Science, 1991). It may even be appropriate for the mentors to follow Adams's (1992) suggestion of grading oneself (e.g., a report card for faculty mentors) as mentors to reflect on what may be strengths and weaknesses. Perhaps, such a review prior to entering the relationship would benefit both the prospective mentor and protégé, especially, since the availability of the mentor is central to a successful relationship (Valadez & Duran, 1991). This can, however, be a major obstacle when the mentor and protégé are not in the same physical location. This can be a particular concern for establishing effective mentoring relationships within the framework of a professional association since personal contact typically occurs at national and/or regional meetings. This does not suggest that this limited opportunity for personal contact will prevent an effective mentoring relationship within the association but rather requires other means of maintaining contact (e.g., telephone, written correspondence, and e-mail). It is clear that much is expected from the mentor. Similarly, the role and responsibilities of the protégé are equally important to the success of the mentoring relationship.

DEFINING THE ROLE OF THE PROTÉGÉ

The mentor plays an instrumental and fundamental role if a mentoring relationship is to be successful. Yet, on her or his own the mentor is still only one side of a two-sided coin and the protégé is the other side. The protégé is the recipient of the mentor's knowledge, wisdom, encouragement, and protection. Although in reality a certain amount of subservience is typically expected to be present, they should not be a passive participant in the mentoring relationship. The protégé should enter the relationship exhibiting a positive attitude that demonstrates a willingness to learn, listen, and act on useful advice offered by the mentor. She or he should also be willing to develop proficiency in certain academic, professional, and scholarly skills that contribute and build upon the mentor's input. The protégé's dependability in completing tasks assigned by the mentor, receptiveness to meaningful criticism, and surpassing the mentor's expectations for performance (Adams, 1992; UCLA Graduate School of Library and Information Science, 1991) can be an incentive that results in stronger advocacy by the mentor. At the same time she or he should feel comfortable expressing concerns to the mentor regarding the progress and direction of the mentoring relationship. Yet, it is also important for the protégé to have realistic expectations regarding the outcomes of the relationship based on their acquired understanding of the mentor, the context of the relationship, and other factors (e.g., race, gender, and politics) that may influence the outcomes of the relationship. Even with this caution for the protégé to be realistic in their expectations of what transpires during the mentoring relationship and the subsequent outcomes, the benefits can potentially outweigh the shortcomings.

As mentioned previously, mentoring has typically occurred within an informal structure. At the same time, this structure is very selective. Not everyone has had a mentor but most generally agree that having one is an asset in the successful attainment of a graduate degree (particularly the doctorate) and in advancing one's career. The dominance of the informal mentoring relationship remains, but there is a recent trend for the establishment of a more formal mentoring structure being explored and implemented by companies, the federal government, universities, and professional associations. Still these efforts can be considered to be in the early stages of development particularly when women and people of color are the targeted populations. While these efforts suggest that the presence of faculty of color are to serve as mentors, at those institutions primarily responsible for training educational researchers, it remains a major concern.

THE IMPACT OF THE SHORTAGE OF FACULTY OF COLOR ON THE POOL OF FUTURE EDUCATIONAL RESEARCHERS OF COLOR

It is not true that all educational researchers have doctorates. However, having a doctorate is very important if one expects to advance his or her career in educational research or academic research in general. The majority of educational researchers are likely to have doctorates. Not having one could be a distinct disadvantage when competing for jobs and career advancement in this area. Not having a mentor during a doctoral program also places one at a distinct disadvantage. In fact, it has been suggested that mentoring is "the heart" of a graduate education (Cusanovich & Gilliand, 1991; as cited in Adams, 1992, p. 1).

Adams (1992) notes that there are four basic stages in the doctoral program where the mentor/doctoral advisor has particular importance.

1. Planning the academic program and course selection.
2. Planning and preparing to take and pass the comprehensive exam.
3. Identifying and conducting an original piece of research.
4. Developing, producing, writing, presenting, and successfully defending the dissertation.

While these are the things that are normally expected of the mentor/doctoral advisor, their level of concern and enthusiasm may vary depending on whether the student has been assigned to them or if the mentoring arrangement was established based on mutual consent. In either scenario the "right" match between the mentor/doctoral advisor and doctoral student/protégé will influence successful progress at each of the above-noted stages in the doctoral program. Doctoral students of color are at a disadvantage when it comes to selecting a mentor or being selected by one.

Whether formal or informal, mentoring is primarily a one-to-one relationship and the mere nature of human interaction makes the mentoring relationship quite complex. This complexity increases significantly at each layer when there are differences based on gender, race, and possibly socioeconomic background. These differences, when combined with the informal relationship and the unwritten rules of the informal mentoring relationship, have negatively affected the success of women and people of color during their doctoral studies, career development, and their pursuit of career advancements. The subjective evaluation and personal biases of the mentor in determining who is worthy of their mentoring is indeed a limitation of mentoring within an informal structure. It is clear that white males have been the primary beneficiaries of successful mentoring in academe (Leon, 1993). Conventional wisdom suggests that mentors are most likely to select protégés who are most like them in terms of academic and professional interests, beliefs, and possibly race and/or gender.

The mentoring needs of the doctoral student of color will not likely be significantly different from that of the majority doctoral student in completing the necessary courses and passing the comprehensive exam. They may seek different elective courses which may address special topics on multicultural issues and take required courses from faculty members who have special interests in these areas and likely incorporate these interests in their courses. However, once the comprehensive exam is passed most doctoral students of color tend to identify research topics that are related to the disadvantaged and/or groups of color. It is at this stage in the doctoral program where the mentor's appreciation for the importance of such topics as credible areas of research inquiry as well as their expertise to contribute to the dissertation research becomes critically important. Not all white faculty are able to meet these expectations and the few faculty of color at research-oriented institutions are overextended as mentors of doctoral students of color.

Atkinson and colleagues (1994) agree that gender, social, and cultural attributes such as race, social class, and religion tend to influence mentors' selection of protégés. These variables also affect the selection of mentors by protégés of color. Unfortunately, the numbers of women and people of color available to serve as mentors continue to be small (Spray, 1994). Consequently, protégés of color have a smaller pool of mentors to draw from even if majority mentors, who value mentoring these students, are included. The numbers are even smaller among the ranks of faculty of color at research-oriented universities. One meaningful way to illustrate this situation within the context of increasing the pool of educational researchers, is by looking at the racial composition of university faculty and doctoral recipients of color in education.

Numerous earlier reports and studies provided clear evidence that faculty of color were grossly underrepresented in colleges and universities across all disciplines (see for example, Astin, 1982; Blackwell, 1987; de La Luz Reyes & Halcon, 1988; Jackson, 1991; Moore, 1988). There has been very little change in this regard. For example, it was reported that for fall 1987 the total percentage of full-time faculty in U.S. four-year institutions included: 5 percent Asian Americans, 3 percent African Americans, 2 percent Hispanics, 1 percent Native Americans, and 89 percent whites (National Center on Postsecondary Education, 1990). By 1991-1992 faculty of color represented 12.3 percent for a modest one percent increase over this four-year period (Adams, 1992). The 1996 Digest of Education Statistics has the following breakdown for full-time faculty in higher education as of fall 1993: 4.6 percent Asian American, 4.7 percent African American, 2.2 percent Hispanic, 0.36 percent Native American, and 85.6 percent white. Still, this does suggest that some progress has been made but a different light may be cast when we consider the presence of educational researchers of color on our faculties and in nonfaculty positions.

The underrepresentation of full-time faculty and educational researchers of color can be in part attributed to the diminished supply of doctoral recipients of color who receive their degrees from research-oriented colleges/schools of education. Faculty of color in the field of education are more heavily concentrated in four-year colleges than in elite research institutions (National Center on Postsecondary Education, 1990). Their absence is believed to result in fewer doctoral students of color seeking to attend elite universities due to what they perceive to be an inhospitable campus climate and less than satisfactory relationships with faculty (Jackson, 1991). This may suggest that prospective doctoral students of color believe that the opportunity to be mentored is restricted by the small number of faculty of color and less than satisfactory mentoring relationships with majority faculty. It is possible for some doctoral students, junior faculty, or new educational researchers of color to have positive mentoring experiences with a majority mentor. Yet it is only realistic to recognize that cross-race mentor-protégé pairs are faced with formidable obstacles that can negatively impact the effectiveness of the mentoring relationship.

Atkinson and colleagues (1991, 1994) reluctantly note that both mentors and protégés of color ratings of mentoring tended to be more positive with ethnically similar or other minority pairings. A majority mentor may have genuine good intentions but their effectiveness in mentoring a protégé of color may be hindered because the expected one-on-one alliance required to exist between them has not been the norm in doctoral programs and may be complicated by the protégé's misgivings, apprehension, and fear (Adams, 1992). Doctoral students of color when faced with few or no prospects for a matched racial pair mentoring relationship may choose a majority member that will at least not deter their progress toward graduation. Settling for a mentor who will not deter the protégé's progress rules out the mentor operationalizing their role as the protégé's protector from the inevitable instances of racism that will be encountered. Grant, Ward, and Forshner (1993) note that doctoral students of color in her study did not see racism as being necessarily blatant in their programs, but they often did not feel supported by their mentors when it did arise. One explanation is that many majority mentors' limited experiences in recognizing subtle forms and instances of racism may prevent them from providing the doctoral student of color support or protection at critical times. The unaware majority mentor may not be sensitive to the psychological toll which instances of racism may have on a doctoral student of color who already feels he or she is in turbulent waters and his or her "lifeguard" only sees calm water. It is understandable that majority mentors who through no fault of their own have little or no experience in being the object of overt or covert racism may be ill-equipped to be an effective protector of the doctoral student of color in carrying out their mentoring responsibilities. While understandable, it is not acceptable. Therefore, exposure to mentors who have shared similar experiences with racism during their doctoral studies and professional careers can be valuable even if they are not faculty of the doctoral program or even their university. The next section of this chapter discusses how professional associations can contribute to a more positive experience during doctoral study for students of color and also aid in their transition into the profession.

MAXIMIZING THE POTENTIAL FOR MENTORING DOCTORAL STUDENTS OF COLOR WITHIN PROFESSIONAL ASSOCIATIONS

Doctoral students and new faculty researchers of color have in fact had fewer opportunities to be mentored in graduate programs as well as in professional associations than their majority counterparts. Professional associations such as the American Educational Research Association (AERA), American Psychological Association (APA), the American Sociological Association (ASA), American Association of Higher Education, and others have great potential for contributing to the increase in the number of faculty researchers of color and the mentoring that will be necessary for their successful career development and advancement. Professional associations can potentially make the following mentoring related contributions. They can:

1. Expand the network of potential mentors of researchers;
2. Expose future researchers of color to a broader base of emerging research on diverse issues facing people of color and other disadvantaged groups;
3. Provide opportunities for personal contact with researchers of color who are investigating issues and questions that may lead to dissertation topics; and
4. Provide opportunities for personal contacts who are willing to serve as resources for employment opportunities, career advancement, letters of recommendation, and letters of support during tenure review.

These are only a few of the possible contributions professional associations can make in the efforts to increase the participation of people of color in research. But they are a start for this discussion.

We have already argued that one of the contributing factors to limited mentoring opportunities for doctoral students and new faculty researchers of color is the small number of faculty researchers of color. Considering that most doctoral programs will have only one or two faculty of color (in rare cases three), professional associations that primarily emphasize research will have larger numbers of faculty researchers of color who are members than can be found on any one campus. Therefore, professional associations can serve as vehicles that expand the pool of potential mentors of color. For example, the American Educational Research Association reported a 1996 total active membership of 23,000. Of those identifying a racial category 5 percent (845) were African American, 6 percent (956) Asian American, and 3.6 percent (587) Hispanic. It should be clarified that these numbers and percentages do include student members but they represent 14 percent of the total membership. It is also important to note that approximately 73 percent of the AERA members reported having a doctorate (Russell, 1996). Therefore, the mere number of potential mentors of color available in the association's membership increases the possibility of establishing racially matched mentoring pairs. The presence of researchers of color among the association's membership is also likely to suggest that they are conducting research and presenting research at association conferences that pertain to diverse issues facing people of color and other disadvantaged groups. Both of these factors can serve to attract prospective protégés of color to the association and the profession.

We have also noted earlier that the rapidly changing demographics in the United States necessitates an increased knowledge and understanding of issues concerning groups of color across disciplines. Researchers of color must play an important role in bringing perspectives that will improve the social, psychological, and educational conditions of disadvantaged groups. In our schools and universities it is clear that many approaches intended to improve the academic progress of students of color have not elicited acceptable results. Those individuals from racially underrepresented groups who seek the doctorate for the purpose of pursuing a career in social science and/or educational research typically have a high level of social consciousness and with a likely special interest in the plight of their particular racial group. Therefore, their exposure to emerging research on racially underrepresented and disadvantaged groups which can be found in research-oriented professional associations could play an important role in expanding their knowledge and understanding of critical research topics, questions, and methodologies. This expanded knowledge and understanding could also assist in the identification of a dissertation topic and research questions. While attending paper presentations and collecting copies of papers presented by researchers of color in general, and particularly those who could serve as mentors, it is often the informal personal interactions at receptions and particularly gatherings of special interest groups within the association that are more fruitful for establishing contacts with potential mentors.

TRENDS IN THE INTERNAL EFFORTS OF PROFESSIONAL ASSOCIATIONS TO ADDRESS EQUITY ISSUES RELATED TO THEIR PROFESSIONALS OF COLOR

Most research-oriented professional associations will have organized groups within the structure that are targeted toward individuals who share research interests relative to a particularly underrepresented group and whose members are largely from that particular group or groups. Within the American Educational Research Association there are organized special interest groups (SIGs) for each of the racially underrepresented groups, women, and those that combine research interest in diverse areas (i.e., SIG/Race, Class, and Gender). Of the 51 American Psychological Association divisions one is for racially underrepresented groups (Division 45 Society for the Psychological Study of Ethnic Minority Issues). The American Sociological Association has 37 sections with four for underrepresented groups (Asian Americans; Latino/a Sociology; Race, Gender, and Class; and Racial and Ethnic Minorities). The paper sessions, symposia, meetings, and informal gatherings sponsored by these special interest groups have and continue to serve as the central place where prospective protégés are likely to establish personal contact with potential mentors who are more likely to share their race and research interests. These special interest groups will typically be very committed and willing to exert considerable energy toward mentoring doctoral students and new faculty researchers while at association conferences. But there must also be mechanisms in place within the formal structure of the association that is conducive to increasing the number of researchers of color who are willing to serve as mentors.

Awarding fellowships and other financial awards to doctoral students from underrepresented groups is a frequently used vehicle by professional associations to support the career development and advancement of these groups. Such efforts are important but insufficient toward achieving the broader goal of increasing the pool of faculty researchers within a profession. The association's structure and policies should also reflect a level of commitment achieving a culturally diverse membership within the association and the profession.

In June 1990 the National Education Association (NEA) and the American Association for University Professors (AAUP) convened a special conference to address the shortage of university faculty of color, problems facing the few that are there, and why affirmative action results have yielded minimal results on the faculties of our universities. This conference resulted in the drafting of a joint statement for consideration and subsequent acceptance by their governing boards. The statement entitled; "Mentoring to Increase Minority Participation in Higher Education: The Faculty's Role," pointed to mentoring as a critical factor in increasing the number of university faculty of color by mentoring doctoral students and junior faculty. Subsequently, "Mentoring Minorities in Higher Education: Passing the Torch," was published by the NEA Standing Committee to provide universities with strategies for successful mentoring programs that were being implemented by some institutions (Leon, 1993). This example of two major associations jointly preparing a policy statement to influence institutional policies would increase the number of faculty researchers of color with mentoring playing an important role. The efforts initiated by the American Educational Research Association (AERA) to make its structure and its efforts more responsive to increasing the presence and participation of educational researchers of color provides another example.

In 1972 AERA adopted four resolutions that were intended to commit AERA to use its resources and influence to promote equity in educational research, secure participation of diverse groups in elected AERA positions, and hold executive officers (president and divisional presidents) accountable for such participation (Russell, 1994). By 1980 AERA's Committee on the Role and Status of Minorities in Educational Research became a standing committee with its purpose being to:

> examine and promote minority participation in the structure and polity of AERA and to promote and improve minority participation in the total sphere of educational research and development" (p. 26).

With this purpose in mind, one of the major activities of the committee is its day-long training workshop and mentoring on publishing strategies with there also being funds available for supporting travel to the annual meeting for students who are presenting their work. A significant effort is further noted in AERA's authorization of $750,000 to fund four doctoral students of color who have successfully defended their dissertation proposals ($10,000 annually). The establishment of AERA's Committee on the Role and Status of Minorities in Educational Research as a standing committee, its purpose and activities do send an important message regarding the association's commitment to make opportunities by committing resources, exerting coordinated efforts to influence institutional policies, and making internal policy and structural changes. The pool of potential mentors of color within the associations' membership, the research they publish and present at association meetings, and the opportunity for extended personal contact between prospective protégés and mentors are major advantages. Unfortunately, it is difficult for mentoring relationships to maintain their intensity beyond the duration of the association meeting, matching of mentoring pairs can be haphazard, time is often too short for the relationship to develop fully, and only those protégés that are able to attend the association meeting are able to benefit. Therefore there is a need for the further refinement of professional associations' mentoring efforts that build upon those things which have been noted above as accomplishments. With this in mind, the next section of this chapter will offer a few initial strategies for consideration by professional associations such as the American Educational Research Association, American Psychological Association, or American Sociological Association. While AERA will be used as the example to frame this discussion, the strategies proposed are believed to be applicable for similar associations.

REFINING STRATEGIES FOR MENTORING FOR PROTÉGÉS OF COLOR WITHIN PROFESSIONAL ASSOCIATIONS

The mentoring of doctoral students and new faculty researchers of color has indeed received increased attention over the past few years. It is not surprising that most faculty researchers of color have played important roles in mentoring these two groups. However, due to the small number of faculty researchers of color to serve as mentors it is also important to recognize the major contributions which majority faculty researchers of color have played in this effort. Therefore, this chapter has not intended to lessen the importance of majority faculty researchers' participation in this mentoring effort. Rather, its intent has been to propose mentoring strategies which can potentially maximize the effectiveness of mentors across racial and gender groups who share the commitment to increase the number and success of faculty researchers of color.

The informal structure of mentoring that exists within most professional associations does have its strengths, but also weaknesses. The often lack of systematic and coordinated utilization of financial and human resources can minimize the broader impact and effectiveness of mentoring protégés of color within professional associations. Fortunately, it appears that professional associations do have structures in place which could be utilized, with some refinement, to establish more structured mentoring efforts by maximizing the associations resources. There are few examples of structured mentoring efforts by professional associations. One that has been implemented by the California Librarians Black Caucus (CLBC) and the UCLA Graduate School of Library and Information Sciences to increase the number of African Americans in the library and information sciences does merit review by other professional associations. The CLBC/UCLA "Mentoring Handbook" is an excellent resource and will serve as a foundation for a number of the strategies which follow for mentoring doctoral students and new faculty researchers of color in professional associations. The professional association that will be utilized to frame this discussion of potential strategies is the American Educational Research Association (AERA).

MAXIMIZING THE ASSOCIATION'S EXISTING STRUCTURE FOR MORE EFFECTIVE MENTORING

As mentioned previously, AERA is a research-oriented professional association which emphasizes the development and dissemination of educational research and has approximately 23,000 members. The association's structure includes 11 divisions that are devoted to broad areas of educational research such as: curriculum studies, measurement and research methodology, learning and instruction, social context of education, and postsecondary education. In addition to the 11 divisions there are also 116 special interest groups (SIGs) that have more highly specialized areas. Members will typically have multiple divisional and SIG affiliations with some type of informal mentoring effort occurring in each entity. It is those SIGs that are devoted to research on particular racial groups (e.g., Research Focus on Black Education) which will place the mentoring of doctoral students and new faculty researchers of color as a major priority without any encouragement from the association but the association's resources and support are important to their efforts. The divisions have more influence and financial resources which could be used to establish more structured mentoring efforts throughout AERA.

One of the first steps in establishing a structured mentoring program within a professional association would be the articulation of the programs goals and objectives. The goals and objectives of a mentoring program for protégés of color are likely to be quite similar across associations with only minor adjustments to make them specific for that particular association. For example, the CLBC/UCLA mentoring program was developed to increase the number of African Americans in the library and information sciences with the major objectives of the program being to increase these students awareness of the profession (as well as career options), encourage their completion of graduate programs in this area, and utilize professionals in the area to serve as mentors. Similarly an AERA division, such as Division D (Research and Measurement Methodology) could have as the goal for its mentoring program to be the increase in the number of educational researchers of color in the area of research and measurement. The objectives to support this goal could be to:

1. Increase the awareness of doctoral students of color regarding career options in research and measurement;
2. Utilize practicing faculty and nonfaculty educational researchers in the area of research and measurement to serve as mentors who supplement the advising received by doctoral students of color within their programs particularly at the post comprehensive examination stages;
3. Utilize practicing faculty and nonfaculty educational researchers in the area of research and measurement to serve as mentors to new doctoral recipients of color in their career development and advancement.

Such statements could be tailor-made for each of the 11 divisions within the AERA structure. However, a critical question would be who will be responsible for developing, implementing, and monitoring the extent to which the division's mentoring program achieves its goals and objectives.

Each AERA division has an affirmative action committee which advises the respective division vice president on affirmative action issues such as membership diversity, nominations for elected division and association positions, liaisons with special interest groups, and mentoring prospective protégés from underrepresented groups. The affirmative action committee seems to be the most logical entity for establishing, implementing, and monitoring the division's mentoring program. It is possible that the division affirmative action committee could be made responsible and supported by the division to serve like the mentoring advisory committee established by the CLBC/UCLA mentoring program. The affirmative action committee would make recommendations to the division vice president regarding appropriate mentoring activities, required resources, budget, and evaluation of the program's effectiveness but would be primarily responsible for overseeing the program. One of the major indicators of the division mentoring program's effectiveness will be how well it facilitates the establishment of mentoring for protégés of color.

FACILITATING THE ESTABLISHMENT OF MENTORING RELATIONSHIPS

The facilitation of establishing mentoring relationships is one of the most critical roles that a professional association's mentoring program can play. It is a start to put potential mentors and protégés in the same place at the same time but it is insufficient. The AERA division affirmative action committees have few examples of consistent and coordinated mentoring efforts targeting prospective protégés of color. However, there have been recent efforts noted in this direction. For example, in 1991 Division D's (Measurement and Research Methodology) Affirmative Action Committee held its first annual mentoring reception for graduate students of color as an opportunity to establish mentoring relationships. The committee quickly recognized that one of the shortcomings of such a reception is that only those graduate students of color who were able to attend the conference would benefit from this activity. Even those doctoral students who were able to attend the national conference one year may not be able to afford to attend the following year. Therefore, the division's affirmative action committee established a mentoring database as a tool to broaden the pool of prospective protégés of color and identify those interested in mentoring this particular group of protégés.

The prospective mentors and protégés completed information forms that included information regarding academic background (degrees completed or in progress), addresses, phone numbers, e-mail addresses, and research interests. Optional race information was also requested. Separate lists of prospective mentors and protégés were prepared that included the above information and then sent to the participants. Prospective protégés were sent the listing of prospective mentors and mentors the listing of the protégés. It was suggested that the protégés review the research interests of the prospective mentors and contact them. The initial group of database participants was relatively small with a combined total of approximately 30-35 prospective protégés and mentors. All of the protégés were doctoral students of color who attended the reception and/or the 1992 AERA Annual Meeting (approximately 20) and the prospective mentors were both white and nonwhite. It is also interesting to note that more than half of the initial group of database participants were not members of Division D. They became participants as a result of attending the mentoring reception, discussions with the chair of the affirmative action committee, and interest since a similar database was not in place elsewhere in the association. The following year increasing numbers of Division D members participated in the database as both mentors and protégés. Some prospective protégés indicated that they had joined Division D because of the mentoring activities but the database also was advertized in the division newsletter and the division vice president called for more members to participate. By 1996 the Division D mentoring database reported more than 90 participants with approximately 40 prospective mentors and 50 protégés of color. While the increase in the number of participants as well as the increased familiarity of the database within the division and AERA its overall impact cannot be fully judged.

Only anecdotal information can be reported regarding the extent to which the doctoral students of color who participated in the database benefitted as a group. There were isolated reports of successful mentoring relationships being established as a result of the database but also failed attempts. An unexpected outcome was that the doctoral students made use of the protégés' listing to contact each other regarding similar research interests and dissertation topics. It is reasonable to suggest that the mentoring database for doctoral students of color was an important start to develop a more structured mentoring effort. But it was only a first step. The next step should be making better use of the database to match mentors and protégés.

One of the limitations with the mentoring database was that not all prospective protégés made use of database. A brief survey developed by the chair of the affirmative action committee and informal conversations with database participants at the AERA annual meeting indicated that some student participants were unsuccessful in their efforts to make contact with the prospective mentors. Other students were more were successful in making contacts. E-mail messages were exchanged, there were telephone conversations, and some reported they had received advice on dissertation topics, methodologies, and reviews of drafts pertaining to the dissertation. Some student participants mentioned that they were a bit intimidated about contacting some of the prospective mentors because of their scholarly reputations and did not call. It seems that if there was a mechanism to facilitate and intervene in the establishment of the mentoring relationship, more benefit could be gained from the mentoring database.

It could be possible for the members of the affirmative action committee to annually review the information on the database participants for the purpose of facilitating protégé-mentor matches. Student participants could be queried via letter or e-mail if their mentoring needs were being met and if not the chair and/or other members of the affirmative action committee could review that student's information and begin by identifying prospective mentors who shared their research interest. Once there was a match of research interest then other factors such as race, gender, and geographic location could be considered for the purpose of identifying possible mentors. The prospective mentors would be contacted first to determine if they were indeed interested and able to serve as a mentor before contacting the prospective protégé. This would prevent the student from being rejected or struggling with how to break the ice at the beginning of the relationship. Once the commitment of the potential mentor had been confirmed the prospective protégé could be contacted and informed of the prospective mentor(s) and then encouraged to contact that person. This would be one way to more effectively utilize the mentoring database. But it would also be equally important for there to be some periodic follow-up from the affirmative action committee regarding the progress and status of the mentoring relationship. All of the above-noted efforts and activities would then be annually reported to the division's membership.

CONCLUSION

This chapter is one more strong voice stating that mentoring is one of the most important ingredients in career development and advancement across professions and academic disciplines. We are reminded that a mentor is not only expected to serve as the protégé's tutor/advisor but also as a protector. Mentors have been entrusted with the responsibility of navigating the turbulent waters of doctoral study and career advancement for those aspiring to become researchers on university faculties or in nonfaculty positions. Professionals of color who have sought careers in research have had limited oppotunities to benefit from sensitive and effective mentoring. The shortage of researchers of color and genuinely concerned majority researchers to serve as mentors has had a negative impact on the need to increase the number of researchers of color across disciplines. Consequently, there is a need to expand the pool of mentors for doctoral students and new faculty researchers of color. Professional associations have the potential to be a valuable resource in this regard.

It is encouraging that professional associations such as the National Education Association, American Association of Higher Education, American Association of University Professors, and the American Educational Research Association have attempted to make themselves more responsive to the goal of achieving a more culturally diverse faculty in our universities. Because of their size, professional associations have an expanded pool of minority and majority researchers who could make a difference as mentors to protégés of color. Some associations have devoted considerable financial resources, refined their internal structures, and taken steps to influence institutional policies in the effort to achieve this diversity. These good intentions are an encouraging start but considerable refinement must occur so that the association's mentoring efforts are better coordinated and systematic so that all of its resources have a maximum impact on the effort to increase the number of researchers of color. There must be an established entity within the association that is responsible for developing, implementing, and monitoring a mentoring program for this targeted population. An affirmative action committe could be such an entity and could play a major role in facilitating the development of productive mentoring relationships. This is only one of several possibilities. The important point is that associations appear ready to be more active players in this effort, have sufficient internal structures to build on, and have growing numbers of researchers of color in their memberships. The question is whether the potential of professional associations will be fully realized in this regard.

REFERENCES

Astin, A. (1982). *Minorities in higher education.* San Francisco: Jossey-Bass.

Adams, H. (1992). *Mentoring: An essential factor in the doctoral process for minority students* (Report No. HE 026497). Notre Dame, IN: National Consortium for Graduate Degrees for Minorities in Engineering (ERIC Document Reproduction Service Number ED 358 769).

Atkinson, D., Casas, A., & Neville, H. (1994). Ethnic minority psychologists: Whom they mentor and benefits they derive from the process. *Journal of Multicultural Counseling and Development, 22*(1), 37-48.

Atkinson, D., Neville, H., & Casas, A., (1991). The mentorship of ethnic minorities in professional psychology. *Professional Psychology Research and Practice, 22*(4), 336-338.

Blackwell, J.E. (1987). *Mainstreaming outsiders: The production of black professionals.* Dix Hills, NY: General Hall, Inc.

del la Luz Reyes, M., & Halcon, J. (1988, August). Racism in academia: The old wolf revisited. *Harvard Education Review*, 299-314.

Epps, E.G. (1989). Academic culture and the minority professor. *Academe, 36* (5), 23-26.

Frierson, H.T., Jr. (1990). Situation of black educational researchers: Continuation of a crisis. *Educational Researcher, 19* (2), 12-17.

Funk and Wagnalls New International Dictionary of the English Language, Comprehensive Edition (1989). New York: World Publishers Guild, Inc.

Grant, L., Ward, K., & Forshner, C. (1993). *Mentoring, gender, and careers of academic scientists* (Report No. SP034679). Atlanta, GA: American Educational Research Association (ERIC Document Reproduction Service Number ED 361 299).

Harris-Schenz, B. (1990). Helping with the bootstraps: The mentor's task. *ADFL Bulletin, 21*(3), 18-21.

Hood, S., & Freeman, D. (1995). Where do sudents of color earn doctorates in education?: The "Top 25" colleges and schools of education. *Journal of Negro Education, 64* (4), 423-436.

Jackson, K. (1991). Black faculty in academia. In P.G. Altbach & K. Lomotey (Eds.), *The racial crisis in American higher education.* Albany, NY: State University of New York.

Leon, D. (1993). *Mentoring in higher education: Passing the torch* (Report No. HE026970). Washington DC: National Education Association (ERIC Document Reproduction Service Number ED 365 195).

National Center of Post Secondary Education (1990). *Faculty in higher education institutions, 1988.* Washington, DC: U.S. Department of Education, Office of Educational Research Improvement.

Moore, W. (1988). Black faculty in white colleges: A dream deferred. *Educational Researcher, 17*, 117-121.

Philip-Jones, L. (1983). Established a formalized mentoring program. *Training and Development, 37*, 38-42.

Russell, W. (1994). Achieving diversity in academe: AERA's Role. *Educational Reseacher, 23*(9), 26-28.

Russell, W. (1996). Annual report: Report in the membership. *Educational Researcher, 25*(6), 36-39.

Spray, K. (1994). *Mentoring: Utilizing this relationship in the graduate education of psychology students and possible concerns* (Report Number CG025861). New Mexico: Eastern New Mexico University (ERIC Document Reproduction Service Number ED 378 466).

University of California Graduate School of Library and Information Science (1991). *California librarians black caucus/ UCLA mentor program handbook* (Report Number IR053963). Los Angeles, CA: University of California (ERIC Document Reproduction Service Number ED 344 591).

Valadez, J.R., & Duran, R. (1991). *Mentoring in higher education.* Paper presented at the annual meeting of the American Educational Research Association, Chicago, IL.

A MENTORING APPROACH INVOLVING HBCU FACULTY'S ON-SITE INVOLVEMENT WITH RESEARCH UNIVERSITY FACULTY

William W. Malloy and Gloria L. Harbin

The pressure on public schools to provide high quality special education has been sustained by federal regulations, state mandates, local district policies, special education advocates, and parent groups and scholars (Fine & Ash, 1988; Gartner & Lipsky, 1987; Lipsky & Gartner, 1989). This pressure has also reached higher education where faculty members are needed in the training and education of teachers and administrators, in addition to the teaching of content related to instruction, school finance and law, and school support services related to special education (Crowson & Boyd, 1993; Simpson, Whelan, & Abel, 1993; Vautour, 1993). Concurrently, there is also the pressure for more African-American faculty members in special education, for as in most areas of academe, the number of African-American faculty in special education programs is woefully underrpresented. Given their small numbers, it is further important that an increased number of African-American special

Diversity in Higher Education, Volume 1, pages 161-174.
Copyright © 1997 by JAI Press Inc.
All rights of reproduction in any form reserved.
ISBN: 0-7623-0086-8

education faculty are more engaged in research that could possibly affect national and state policies associated with special education. Historically black colleges and universities (HBCUs) are a source to engage more African-American faculty in increased research efforts.

There are currently 101 HBCUs in the United States. Of these, approximately 90 percent have either a special education department or program. These institutions are responsible for training and educating a sizable portion of African-American special education teachers in the United States. For example, in North Carolina there are nine HBCUs with special education training programs that train and graduate approximately 80 percent of the African-American special education teachers in the state. Thus, it is clear that the nation's HBCUs play a major role in preparing African-American special education teachers for our public school systems. Given the influence that these institutions exert on their local communities, according to Roebuck and Marty (1993), the presence of African-American college faculty in special education departments or programs directly or indirectly affect recruitment of African Americans and other individuals of color into special education teaching and administration areas. Moreover, as indicated, it is important for HBCU special education faculty to engage in research that may possibly affect policy and to expand the knowledge production related to the field of special education. Toward addressing the need for increasing the number of underrepresented minorities in special education and the engagement of those faculty in more research, the University of North Carolina decided to initiate a mentoring program involving HBCU faculty that would lead to the increase of minorities in special education areas related to research and teaching at all levels. This mentoring program was implemented through a postdoctoral approach.

THE POSTDOCTORAL MENTORING PROGRAM

In 1992 the School of Education at UNC-CH began exploring the possibility of collaborative strategies with HBCU special education departments. This exploration resulted in the development of five goals for collaborative strategies.

1. To strengthen the knowledge of the postdoctoral fellow in selected content areas related to special education.
2. To strengthen the knowledge and leadership ability of the postdoctoral fellow so that he or she will be able to enter and influence various subsystems once he or she has returned to the home department, thus ensuring that what was learned in the postdoctoral experience becomes integrated into the various appropriate subsystems.
3. To increase the capacity of the university training programs at the HBCUs to provide quality training and research.

4. To enhance the expertise of the individual to increase their ability to compete for and successfully obtain federal and state grants.
5. To enhance the professional's network by offering them the opportunity to collaborate with highly recognized scholars in the field of special education.

The strategy most favored was a postdoctoral program approach that would be personally rewarding for the scholar while simultaneously benefitting the scholar's home campus's special education department. To address both aims, the program was designed to encourage a high degree of flexibility but sufficiently structured to ensure that end-product would result. This flexibility would enhance the scholars' efforts to pursue a research agenda, publish, and/or successfully compete for grants. In return for this flexibility, the scholar would be required to initiate one project that would be transferred to the home campus. The project could include writing a proposal for funding a personal preparation initiative, designing new components for current special education teacher training programs, or developing strategies for interdepartmental or campus/community collaboration related to special education. All activities would focus on increasing the visibility and efficacy of special education on HBCU campuses and thus improve the enrollment of more students in special education programs.

The Admissions Process for the Program

Over a five-year period eight fellows have participated in the program. This group was composed of one tenured full professor, four tenured associate professors, two tenure-track assistant professors, and one recent Ph.D. recipient. All fellows were African American and two of the six fellows were males. A total of 10 mentors served the fellows. Of the mentors, seven were tenured full professors, two were associate professors (one was tenured and the other was a clinical associate professor—fixed termed), and one was a tenure-track assistant professor. Two mentors were African Americans and the other eight were white. Regarding the mentors' gender, five were female and five were male.

A rigorous admissions process ensured that only individuals with a sense of purpose related to the program's goals were enrolled. Each eligible applicant must have a completed doctoral degree and are required to submit the following: (1) a complete resume or curriculum vita; (2) three letters of recommendation; (3) a detailed rationale describing the individual's goals for the postdoctoral experiences; (4) how these goals and experiences relate to the professional and leadership goals of the applicant; and (5) a personal statement of why the applicant feels those goals could be achieved at UNC-CH. An admissions committee reviews each application and evaluates the candidate's potential for successful participation in the program. In the case of applicants of relatively equal quality, efforts are made to make decisions that can result in a group of fellows that are diverse with respect to areas of interest, home institution, and gender.

Our aim is for our postdoctoral fellows to be a diverse group and, further, we expect that each will be able to take leaves from their respective institutions. Experience has shown that some faculty candidates may be entitled to a sabbatical and would be able to spend a full year on-site at UNC-CH. However, many private HBCUs do not have the economic resources to grant sabbatical leaves to junior faculty members, and those are the individuals, according to a survey we conducted, who are most interested in the postdoctoral experience. In addition, experience and survey findings indicate that, typically, those junior faculty members interested in the postdoctoral program are the major or only source of income for their families, and are unable to afford a year-long unpaid leave. As a result of these considerations, we offered the postdoctoral fellow the choice of three options to facilitate smooth functioning within the fellow's department:

Option 1: Full-time, on-campus at UNC-CH for one full year
Option 2: Two consecutive semesters (e.g., spring and summer or summer and fall)
Option 3: Two summers.

Thus, participating fellows can have time options of ranging from four (two consecutive summers) to 12 months. These options allow prospective fellows the opportunity to participate within a time frame that is commensurate with their needs and the boundaries allowed for participation.

Financial Support

We had originally anticipated, during the five years of program operation, that the postdoctoral fellows would receive partial salary support from their universities while participating in the fellowship program. We provided a $17,500 stipend for each of the projected three fellows per year or $52,500 in total annual fellowship stipends. Experience demonstrated that this was not a sufficient stipend because the majority of the sending institutions could not provide partial salary support at that level.

As a result of our previous experiences, we now provide a minimum fellowship stipend of $17,500 for those fellows fortunate enough to receive partial salary support. In other instances, fellows unable to receive partial salary support and experience financial hardships, may receive a stipend up to a maximum of $36,000. An annual total of $60,000 for fellowship stipends ensured that at least two fellows would receive sufficient funds to participate in the program. However, in some years, depending upon the financial resources of the sending institutions and the personal finances of the potential fellows, we fully expect three fellows to participate in the program.

Expectations of Fellows

Informative Interview

Once the postdoctoral fellow has been selected, he or she comes to campus for an initial interview with the program's project coordinator. The purpose of these initial interviews is to obtain information that will be beneficial to the eventual development of a joint plan for the postdoctoral experience involving the University of North Carolina at Chapel Hill and the prospective fellow's home campus. The following information is collected from the participating fellows during the interview:

a. What are their areas of training experience?
b. What do they teach?
c. Do they have other professional responsibilities?
d. What are their professional goals?
e. In what content areas would they like to focus?
f. What do they want to gain from this experience?
g. Does their department/university have a long-range plan?
h. How might this opportunity enable them to not only be better prepared to implement the plan, but to provide leadership within the department?
i. In which professional arenas are they interested in exercising leadership (e.g., public schools, schools systems, colleges/universities, and funding agencies)?

MENTORSHIP ASSIGNMENT PROCESS

Orientation

The postdoctoral fellows receive an orientation through a series of presentations and written materials related to ongoing projects and possible experiences within each of the project and content areas. During a one-day orientation based upon content areas identified in previous interviews with postdoctoral fellows, selected faculty members from the University of North Carolina at Chapel Hill make presentations about their projects. There is an informal reception to which a wide array of faculty are invited. This gives fellows the opportunity to meet a wider group of faculty members and to do so in a more informal setting while having the opportunity to gain greater knowledge about the research in which various faculty are engaged.

Selection of Focus Areas

After the orientation the postdoctoral fellows will then have a meeting with the project staff to discuss their reactions to the faculty presentations. Additionally, fellows have the opportunity to discuss the various array of special education content areas, as well as how concentrating on any of these areas might be of benefit to furthering their professional goals and leadership abilities. In this meeting postdoctoral fellows finalize the selection of the content areas off campus for their postdoctoral experience. They also select a faculty partner (mentor) who is instrumental in helping to develop and implement the fellows' individual plans or learning programs.

Develop Individual Postdoctoral Plans

The postdoctoral fellow meets with the selected faculty partners, and through a process of conversations, selects the UNC-CH experiences and projects in which they wish to be involved. As part of the application process, each postdoctoral fellow develops a set of objectives for his or her postdoctoral experience. As a result of discussions with the project staff and the faculty partner, the postdoctoral fellow has the option of revising his or her objectives. With assistance from project staff and the faculty partner a plan of activities and experiences are developed to meet any revised objectives. The objectives for the plan focus on the improvement of one or more of the following skills: university teaching, research, in-service functions, and writing. This plan includes the identification of the product (e.g., a new course) that is to be developed once the fellow returns to his home department, as well as the type of grant (research, training, etc.) to be developed during the postdoctoral experience. The project staff (known as the system-based support team) and the faculty partner (i.e., the mentor) is responsible for monitoring and evaluating the success of the postdoctoral experience.

Design and Development of Product and Grant Proposal

At some point early in the postdoctoral experience, the fellow, through consultation with the mentor, designs the product and the grant. Throughout the product development process the fellow has opportunities to consult with his or her mentor and the project staff. The mentor, as well as project staff, also facilitate access to other faculty and resources across the university, as well as in the School of Education and the Frank Porter Graham Child Development Center as needed or desired. These products and grant proposals are meant to be used to enhance the capacity of the postdoctoral fellow's home institution once he or she returns to that site.

Presentation

Each postdoctoral fellow makes a presentation to UNC-CH students on a topic of his or her choice. The topic could be one that is familiar to the fellow or something in the area of concentration that was selected for the postdoctoral experience. The latter would be preferred since it would reinforce activities occuring during the fellowship period. The presentation has two purposes. First, it serves as a vehicle for the postdoctoral fellow to receive feedback on both the content and method of the presentation. Both the faculty partner and the project staff may provide feedback to the fellow. Second, the presentation serves to put individuals of color in teaching/leadership roles for UNC-CH students. As in most predominately white universities, individuals of color are severely underrepresented in the faculty ranks. The opportunity to experience an individual of color in teaching roles is beneficial to both students from predominant and underrepresented groups.

Monthly Meetings

The director of the postdoctoral program meets monthly with the fellows and mentors to discuss barriers, opportunities, and successes in accomplishing their goals. These meetings serve as evaluative functions and to ensure that the fellow is gaining the experiences desired.

FOLLOW-UP POSTDOCTORAL EXPERIENCES

The nature, the amount, and the person(s) providing the follow-up experiences vary and are primarily based on the individual needs and situations of the fellows. Follow-up activities are designed to: (1) assist the fellow in integrating his newly acquired skills and knowledge into his home institution's department; (2) assist the fellow in gaining access to other special education or professional arenas (e.g., schools, state agencies, the broader university, etc.) and demonstrating leadership; and (3) continue to enhance the capacity of the home institution in quality training and acquiring resources.

Follow-up may be provided by the program director, the mentor, or a combination of both. The form of the support includes on-site consulting on the campus of the fellow's home institution, writing letters of support, making telephone calls to introduce the fellow to important and influential individuals, telephone consultation with the fellow or his or her department chair or dean, reviewing journal article drafts, or recommending to journal editors that the fellow is added to the group of reviewers. These are some examples of the methods of follow-up that are used depending, once again, upon individual situations. Members of the systems-based support team are also willing to make a presentation to a class or perhaps to teach a class session with the fellow after the fellowship period. This would give greater visibility to the fellow, and thus enhance his standing in his department or school. This particular circumstance also provides an opportunity for the UNC-CH School of Education to recruit potential graduate students for its special education graduate program. UNC-CH is the only university in the University of North Carolina system that has a doctoral program in special education.

Fellows are invited back to campus each year to take part in the Visiting Scholars Seminar Series. These seminars provide an opportunity for the fellows to discuss difficulties in integrating skills and knowledge back into their home institution as well as discussing barriers related to efforts at attempts to influence policy and systems. In addition, it provides the opportunity to develop a support network of past and present fellows, and a vehicle for maintaining professional relationships. Whenever it is possible, the Visiting Scholars Seminar is conducted cooperatively by a former fellow and one the project staff.

All of the follow-up activities conducted by the faculty partner and members of the system-based support team are designed to increase the leadership capacities of the postdoctoral fellows, encouraging them to act as leaders at multiple levels of the professional ecological system.

RESULTS OF UNC-CH'S POSTDOCTORAL PROGRAM

A recent survey (1995) of African-American special education faculty in selected HBCUs indicated general interest in a postdoctoral program. There is evidence that such programs are needed and there is also interest in such opportunities (Brazziel, 1987; Brown, 1985: Federal Regional Resource Center, 1991). Recruitment for this postdoctoral program over a three-year period has met with uneven success. That study was conducted as a strategy for identifying possible program revisions to improve the recruitment of prospective fellows. In the study special education faculty members responded to questions related to their perceptions of the personal and institutional benefits of the postdoctoral fellowship experience. The views and perspectives of these selected special education faculty at HBCUs regarding the advantages and disadvantages of postdoctoral experiences are discussed in the following section. The findings from this study formed the basis for recommendations used to modify recruitment incentives to exert greater appeal both for HBCUs and for faculty of color at HBCUs.

ADVANTAGES AND DISADVANTAGES TO POSTDOCTORAL PROGRAM PARTICIPANTS

The results from the aforementioned survey revealed that the postdoctoral experience would provide excellent opportunities for participants' educational and professional reinvigoration. The heavy teaching load of faculty at HBCUs and accompanying community commitments have tended to retard their efforts to stay abreast of the field and involved in research. African-American professors at HBCUs, when not teaching, are expected to devote a respectable amount of time involving themselves in community and other service-related activities. Comparative statistics between HBCUs and their counterparts at research I or II universities were not included in this study, but the double commitment that results in significantly heavy teaching and service loads appears to be a common experience which is perhaps unique to HBCU campuses. Notably, Roebuck and Marty (1993) assert, "The Black Faculty member at an HBCU who opts exclusively for research and publications at the expense of teaching and service, runs a risk of incurring negative sanctions for not being a good citizen and not living up to the community service expectations of minority students, peers, and administration" (p. 18).

According to the survey results, using the postdoctoral experience to remain current in one's field was not necessarily linked with an unceasing desire to publish as part of a quest for promotion. This is not a startling revelation since requirements for tenure and promotion at teaching institutions are not overwhelmingly weighted in favor of a heavy record of scholarly publications. However, for several survey respondents, the issue of publications loomed large as a strategy to increase employability potential at non-HBCU research institutions. In fact, one of the previous postdoctoral fellows who eventually transferred to such a research institution perceived that the postdoctoral program had facilitated her career move. It does bear pointing out, however, that the purpose of the postdoctoral fellowship program at UNC-CH is to see the postdoctoral fellows leave their institutions to seek employment at colleges and universities that are more research oriented. The actual intent is to expand the individuals' skills and knowledge so that they can be more effective at their home institution.

A double concern—which is financial and emotional—centered around the absence from family. Considering that most respondents held the rank of senior faculty, this is not a finding of breakthrough proportion. A financial concern for the majority of respondents who considered themselves the family's major financial contributors was that a year-long absence from this role would adversely affect their family's financial well-being. In instances where school-age children were involved, particularly in single-parent situations, some respondents felt that an extended absence might prove harmful to the well-being of the family structure. More specifically, issues of child care for single parents increased the likelihood that participation in the program would not be feasible from both a financial and emotional standpoint. In situations of two-parent families, the extended absence might prove to be an unbearable hardship on the remaining parent with increased responsibilities for financial support competing with quality time spent with children.

PERCEIVED ADVANTAGES AND DISADVANTAGES OF POSTDOCTORAL PROGRAMS FOR HBCU FACULTY

Several respondents cited benefits from the postdoctoral experiences for HBCUs that would result in recognition of their respective campuses. The belief was that presentations or invited speeches at prestigious conventions and professional societies to share data collected during their faculty's postdoctoral experience would surely be forthcoming. Fellowship experiences that might lead to an editorship or co-editorship of a professional journal were also a likelihood and viewed as valuable. Additional views were that any form of national recognition of resident scholars at HBCUs attests to the quality of the individual school's educational program and recognition of resident scholars can facilitate greater opportunities to secure external funds. Therefore, one of the most significant perceived benefits to the schools are fellowship experiences that focus on seeking and obtaining external funds through grants. It was clear that grants would not only significantly increase an individual faculty member's chances of obtaining an official leave from the home campus but also, because colleges and universities thrive on capturing external funds, success in gaining funding grants would further enhance the reputation of the institution itself.

Survey respondents combined their personal concerns with concerns for their institutions. Most respondents felt strong loyalty to their home campus and believed a one-year hiatus might disrupt academic flow. More specifically, their concerns related to the possibility of having adjuncts assume their teaching loads with minimal knowledge or facility with course content. This commitment to the students is an apparent result from the family-like atmosphere that prevails at HBCUs. As observed by Roebuck and Marty (1993) that atmosphere encourages strong mentoring relationships between teachers and students.

ADDITIONAL RESOURCES NEEDED FOR PARTICIPANTS AND HBCUs

The number one need, agreed upon by all respondents, was adequate financial resources. (That need, of course, is of paramount importance for faculty from non-HBCU institutions as well.) The postdoctoral experience is generally perceived as rewarding if the potential components—the sponsoring university, the participating school (in this case, HBCUs), and the participating fellows do not have to suffer undue financial hardship. Unmarried junior faculty respondents with no children were less likely to perceive the experience as one who might strain his or her financial resources. On the other hand, respondents who were married and/or had children viewed such experiences as possibly exerting a considerable financial strain on themselves. In general, if the level of financial support is perceived as adequate, then the other areas of support would take on less significance.

All respondents felt that the other areas of support—those related to clerical services, travel, graduate student assistance, and computer access—were needed; but the absence of one or more would not influence the decision to participate in a postdoctoral fellowship program. Significantly, the desire to secure those areas of support tended to decrease in relation to the size of the fellowship stipend. The next most important area of support cited was related to the availability of highly recognized scholars with whom to collaborate in the form of team-teaching or consultation in specific areas of research. Some respondents felt this could greatly enhance their ability to network with colleagues at both HBCUs and non-HBCUs. Another perceived benefit of this collaboration is the possible increased opportunity to secure external funds. The importance of securing external funding is taking on an ever-increasing magnitude and collaborations between HBCU faculty fellows and with faculty at the research I school appears to be viewed as activities that could result in successful proposals for external funds for HBCU schools and, of course, the faculty fellows themselves.

The findings from the survey provide valuable insight into how a postdoctoral program might be structured to attract scholars from HBCUs. Each postdoctoral program should have a specific focus that is peculiar to that particular program; however, there appear to be some general considerations that might increase the recruitment power needed to attract more faculty members of color, particularly those from HBCUs. What follows are several recommendations that should be considered when initiating or restructuring postdoctoral programs to accommodate specific needs of minority scholars from HBCUs. Faculty from HBCUs, and other institutions that do not have a research mission as its primary focus, considering taking leaves for postdoctoral experiences, might use these recommendations to evaluate current postdoctoral opportunities in relation to their particular needs.

1. Flexible Scale for Stipends. The stipends should be on a sliding scale to accommodate individual circumstances. Generally fellowships are offered without regard to family circumstances, a practice that tends to depress interest. If the funding source requires a uniform level of support, then opportunities

to earn additional income should be made available. The source of the additional funds might be short-term consulting arrangements, honoraria for a limited number of speaking engagements, and compensation for adjunct teaching at local higher education institutions.

2. Emphasis on Grant Writing. As most institutions of higher education welcome external funds, there should be opportunities to write proposals for external funding. Postdoctoral programs that feature proposal writing opportunities might improve recruitment efforts of minority scholars and encouragement for participation from their institutions.

3. Structured Collaboration. Opportunities for fellows to engage in collaborative research would be an appealing attribute of many postdoctoral programs. Pairing fellows with notable scholars in mutual areas of interest might result not only in published papers but in securing external funds as a result of those partnerships. At the very least, this collaboration would expand the professional network of the fellow, a benefit that should prove fruitful in the future. Collaborative experiences that improve the national reputation of the fellow would also be attractive to HBCUs and other schools if they increased recognition of those institutions' faculties.

4. Publications. Greater emphasis would be placed on using the postdoctoral experience to collect data for a book or on actually writing a book, in instances where fellows intend to return to the home campus. Publishing in professional journals and establishing a foundation to build a publication record might receive equal emphasis for those fellows who might eventually be interested in exploring employment opportunities at other universities, or in strengthening a portfolio needed for promotion at the home campus.

5. Institutional Support. There should be an external postdoctoral program that enables the scholar to remain on his or her home campus during the academic year and participate in the postdoctoral program during successive summers. This arrangement might eliminate the economic and emotional concerns associated with extended absences. Additionally, the program might underwrite the cost of released time for the fellow so that his or her institution could provide suitable replacements while the fellow remained on institutional salary. One approach idea might be a postdoctoral experience that permits the fellow to remain on his or her campus with a reduced class load. A notable scholar who would serve as a mentor would periodically visit the campus, expenses underwritten by the program, to engage in collaborative research or to consult with the fellow on research he or she is conducting or on the development of grant proposals. The visit phase of this situation could also be reversed whereby the fellow could visit the campus of the mentor. In either case, the end result might be publications or grant awards.

HBCUs that bestow their blessings and support on scholars who participate in these postdoctoral programs may receive returns on that support. The returns could be in the form of increased external funding support, greater recognition generated from scholarly works of faculty who participated in the program, increased research productivity from those faculty, expanded networks of academic collaboration, and improved student instruction as result of expanded instructional materials that can be developed as a result of participation in the programs.

Attracting institutional support of postdoctoral programs will require that the program be structured in a manner where the institutional benefits are obvious. Postdoctoral programs should provide sufficient support to override personal and institutional impediments that may hurt recruitment efforts. Perhaps the wisest strategy would be to restructure the postdoctoral program as cooperative ventures between participating HBCUs and the sponsoring university, and allowing each HBCU to declare its objective, identify and represent its candidate(s), and delineate the needs to be met for both the institution and the candidate(s). The end results of such restructuring, however, might be to change the perception of benefactor/recipient, and this change, though subtle, might break through the barriers that have hampered the expanded participation of HBCU faculty's participation in postdoctoral programs related to special education.

CONCLUSION

Properly developed and structured, postdoctoral programs can prove to be excellent vehicles for the mentoring process. As a result of the current program's emphasis on mentoring, the postdoctoral fellows were engaged in a wide variety of activities. For some, it helped shape their future research agenda, for others it expanded their repertoire of instructional techniques, honed their proposal writing skills, and/or provided direction for a future book. Of particular note was the increased exposure of the fellows, through mentorships, to an expanded network of academicians with whom to pursue future research initiatives.

Individual perspectives of the eight fellows varied with regards to the perceived value of the postdoctoral mentoring relationships. The two assistant professors found the relationships rewarding as a mechanism for learning more about the political intricacies of the professorship without the scrutiny that accompanies on-the-job performance. In contrast, the associate professors were most encouraged and stimulated by the expanded network of professional colleagues that was facilitated by the mentors.

Importantly, the value of the postdoctoral program for the mentors was immeasurable. The mentors benefitted greatly from the cross-culture perspectives that the fellows shared. Additionally, the mentors became more cognizant of the potential for a variety of conjoint research initiatives between HBCUs and traditional research universities.

REFERENCES

Brazziel, W. (1987). Road blocks to graduate school: Black Americans are not achieving parity. *Educational Record, 68* (4) & *69* (1), 108-115.

Brown, P.A. (1985). *Helping minority students succeed: A report of the Association of American Colleges' minority achievement program, 1981-1985*. Washington, DC: Association of American Colleges.

Crowson, C., & Boyd, A. (1993, February). Coordinated services for children: Designing acts for storms and seas unknown. *American Journal of Education, 101* (2).

Federal Regional Resource Center (1991). *Exploring education issues of cultural diversity*. Lexington, KY: Author.

Fine, M., & Asch, A. (1988). Disability beyond stigma: Social interaction discrimination and Activism. *Journal of Social Issues, 44* (1), 3-22.

Gartner, A., & Lipsky, D.K. (1987). Beyond special education: Toward a quality system for all students. *Harvard Educational Review, 57*, 367-395.

Lipsky, D.K., & Gartner, A. (1989). *Beyond separate education: Quality education for all.* Baltimore, MD: Brookes.

Roebuck, J.B., & Marty, K. (1993). *Historically black colleges and universities: Their place in American higher education*. Westport, CT: Praegar.

Simpson, R., Whelan, J., & Zabel, H. (1993, March/April). Special education personnel preparation in the 21st Century: Issues and strategies. *Remedial and Special Education, 14* (2), 7-22.

Vautour, J. (1993, March/April). Personnel preparation in the 21st century: Response to Simpson Whelan and Zabel. *Remedial and Special Education, 14* (2), 23-24.

PART III

COUNTERVAILING FACTORS RELATIVE TO THE MENTORING EXPERIENCES OF SCHOLARS OF COLOR

NEGATIVE MENTORING: AN EXAMINATION OF THE PHENOMENON AS IT AFFECTS MINORITY STUDENTS

Reginald Wilson

As is pointed out in other chapters of this volume, Mentor, according to Greek mythology was "a wise and trusted counselor and teacher" to his protégé, Telemachus, and we have used the term "mentor" in that sense ever since. A good mentor is a role model exemplifying those characteristics which the protégé wishes to emulate. Additionally, a mentor is a teacher instructing the protégé in the highest possible skills that person can attain. Further, a mentor is a counselor advising the student when straying from the right track and praising when not. Those roles, of course, are consonant with those described by Morris Zelditch in this volume.

Nearly every study of academic mentoring confirms its positive effects with regard to protégés' achievements, self-concepts, and aspirations. The conventional wisdom is that mentoring can only be desirable for students in

Diversity in Higher Education, Volume 1, pages 177-185.
Copyright © 1997 by JAI Press Inc.
All rights of reproduction in any form reserved.
ISBN: 0-7623-0086-8

postsecondary education settings. Thus, for example, De Vos (1973) can speak casually of a maternal mentor who "teaches her son self-reliance and personal competition with a standard of excellence.... [A]s an adult he will continue to seek out tasks that will express these now internalized standards." Likewise, it is thus assumed that mentoring is always positive and the adoption of the mentor's behavior will always lead to greater striving for excellence on the part of the protégé.

A review of the literature indicates some common themes that run through the definition of mentoring. Those themes indicate that "mentoring is a process within a contextual setting; involves a relationship of a more knowledgeable individual with a less experienced individual; provides professional networking, counseling, guiding, instructing, modeling, and sponsoring; is a developmental mechanism (personal, professional, and psychological); is a socialization and reciprocal relationship; and provides an identity transformation for both mentor and mentee" (Galbraith & Coehn, 1995).

All of these definitions, either implicitly or explicitly, suggest that mentoring is a positive experience with consequent growth and development for the protégé. However, I want to remark on the phenomenon of negative mentoring, particularly as it affects minority student achievement, self-concept, and aspiration. As minority students become an increasing part of the national student body, more attention must be paid to this phenomenon.

Negative mentoring can be characterized as consciously or unconsciously: (1) lowering expectations for minority student achievement in the classroom; (2) advising minorities to consider lower majors than their ability calls for; and (3) reinforcing already low self-concept rather than challenging students to aim higher. The literature is sparse in noting any studies of negative mentoring, but rich in qualitative and anecdotal studies of the phenomenon. For example, the story is told by Dr. Raymond Bowen, African-American President of LaGuardia Community College, that his "interest [in science] was almost shattered by my high school biology teacher who told me that I would do better with a shoeshine box. He said, 'There are no Negro scientists.' However, I knew better because my parents taught me about George Washington Carver and Benjamin Banneker" (Bowen & Muller, 1996).

Of course, in reviewing the literature on mentoring, one comes across instances of mentoring that did not work out. Braun (1980), for example, describes a systematic study of poor mentoring relationships that resulted in "mentor's loss of power resulting in diminished career possibilities for the protégé, mentor's destructive personality, dependency and overdependency issues, traditional views of the mentor clashing with the protégé's newfound roles, exploitative relationships, encountering the 'queen bee syndrome,' and possible gender difficulties." While these are certainly real and quite signficant barriers to fruitful mentoring relationships, they are not ones that are the particular focus of this chapter. In studying the mentoring relationship I am especially looking at those mentors who are very positive in their attitude toward the protégé, offering what they believe is advice in the best interest of the protégé, but ultimately are destructive and diminishing of the protégé's capabilities and ambition. Moreover, this phenomenon is observed most often in cases of majority mentors with racial minority protégés.

In this chapter I wish to explore particularly three settings in which negative mentoring has manifested itself: desegregation of schools, the "Pygmalion" effect, and counseling for higher education.

NEGATIVE MENTORING IN SCHOOL DESEGREGATION

School desegregation, despite its historic success in overturning legally sanctioned separation of the races throughout its 45-year history, has been plagued with resistance and volatile feelings of anger on the part of whites which have usually been expressed in negative attitudes toward black students. For example Willie (1976) states that in the Boston Public Schools (one of the most rancorous desegregation sites) "[desegregation] seemed acceptable as long as it did not require a rearrangement of the educational opportunity system to be adequate with reference to the social classes." That is, school desegregation must not alter social class relations, which by its very nature it was designed to do.

But more particularly, with regard to our subject, desegregation often provided a forum for *desegregation* in schools but *segregation* in classes, which give minority children a feeling of inferiority and "far less conviction than whites that they can affect their own environments and futures" (Willie, 1976). Specifically, when placed in classes with white students, black youngsters are often assumed to be inferior in academic achievement, problematic in socialization, and less able to cope with rigorous course material. Thus, they are often given less challenging material, are tracked in lower ability classes, and are separated from the high achieving white students (Rubin, 1976). This will, of course, assure their lower academic achievement and lessen the likelihood of their selecting challenging careers. But this is the finding of poor school desegregation cases of which, alas, they are mostly in the majority. However, the results of *good* school desegregation are rich and rewarding: "those black adults who attended integrated schools were more likely to complete high school, to attend college, to earn somewhat more money, to enter higher status occupations, to have stable marriages, to have steady jobs, to live in integrated neighborhoods, [and] to have white friends and acquaintances" (Crain, 1976).

Similarly, when bad or negative mentoring occurs it tends to lower the minority students' aspirations, dampen academic achievement, and decrease the likelihood that the student will choose a nontraditional or rigorous course of study. Whereas positive mentoring will result in all the good outcomes that have been mentioned as coming from a positive experience of desegregation.

THE "PYGMALION" EFFECT IN MENTORING

This concept is a variation of the Pygmalion effect first dramatically illustrated by Rosenthal and Jacobson (1968). Although the findings of the study aroused considerable controversy at the time, the study was very straightforward. Two groups of students were matched for academic ability and introduced to different classrooms. The teacher of one class was told that her students were above average in ability and could be expected to do very well. The teacher of the other class was told that her students were below average and could be expected to do poorly.

In a classic instance of the self-fulfilling prophecy, of course, the "bright" students did well and the "slow" students did poorly. Moreover, in observing the teacher of the "bright" students, she did all those things that would guarantee her students success: spending extra time, giving strong encouragement, and giving extra academic help. The teacher of the "slow" students did nothing additional to help her students and gave no extra time.

So the "bright" students were not only expected to do well, their teacher did everything to make her expectations come true. As expected, the Rosenthal and Jacobson study was criticized on several grounds, the most serious being the ethics of telling the teachers an untruth. But the results are, nonetheless, quite dramatic and illustrative of the role of expectation on behavior.

In mentoring, as in teaching, the expectation is that all mentoring is positive and reinforcing. As illustrated in the Rosenthal study, and corroborated later by Ray Rist (1970), negative teaching—as with negative mentoring—need not be racist in motivation or stem from a dislike of the student, but the teacher, steeped in the societal expectations for the minority student because of his race or class status, may press the student to lower his career ambitions or suppress his academic expectations.

The teacher might very well believe that he is acting out of kindness toward the minority student by not placing him in "too hard" a course or counseling him away from a "too difficult" curriculum, rather than challenging the student to reach high or to try something different, and providing the student with the encouragement and tutoring needed to succeed in those situations.

COUNSELING AS AN EXAMPLE OF NEGATIVE MENTORING

"Throughout the history of the African-American experience in America, white Americans have expressed doubts about the intellectual capabilities of people of African descent" (Allen, Epps, & Hanniff, 1991). Therefore, often classroom expectations of achievement were lower for blacks and, in a perverse twist of liberalism, classroom achievement was marked high for marginal achievement because "that was the best you could expect of them." As a result, African Americans experienced achievement "shock" when confronted with rigorous academic challenges upon entering higher education. There is the actual case of the black Detroit high school valedictorian failing his first year at the University of Michigan, not because he was not bright enough, but because the classroom achievement expectations had been so much lower for him in his segregated all-black high school. Thus, lowered expectations in high school was a disservice to this student because his high grades gave him false expectations about the achievement level required in college.

Nettles (Allen, Epps, & Hanniff, 1991), in his quantitative study of differences between black and white student college achievement, stressed "the characteristics that students develop before entering college—mainly, their high school preparation," as being most important. Of course, he recognized the importance of other factors—socioeconomic status, college experiences, and so on—but believes the primacy of precollege experience was most important in shaping the expectations of college. Nettles goes on to observe that "black students' high school and college experiences contribute to their college GPAs."

Gail Thomas (Allen, Epps, & Hanniff, 1991) notes in her study of college-major selection by black students that, "Blacks remain highly underrepresented in college majors and careers that offer the greatest opportunity for future job access and earnings." A series of studies document that black students are vastly underrepresented in the natural and technical sciences (Herzog, 1982; Rosenfeld, 1980); and that the income differential between blacks and whites is largely attributable to differences in major field and career choices. Contributing factors to choice of college major are parents' education and high school courses taken. Sells (1976) finds blacks are most likely to come from first-generation college-going families and, thus, make lower career choices; and that blacks take fewer mathematics and science courses in high school, making them ineligible for mathematics and science majors in college.

It is certainly the case that coming from noncollege-going families will depress choosing highly competitive majors, but more directly for our purposes, is the finding that attending a black college (and being influenced by black counselors and teachers) will raise the aspirations and significantly influence more students to aspire to mathematics and science majors. Thomas's study (Allen, Epps, & Hanniff, 1991) corroborates the impact of attending a black college in finding that blacks selected technical and science majors were "60 percent in black private colleges and 42 percent in predominantly white colleges."

A classical example of negative and positive mentoring is one reported by Dr. Antoine Garibaldi, formerly the Vice President of Academic Affairs, Xavier University (a historically black school) in New Orleans, Louisiana and now the Provost and Chief Academic Officer at Howard University in

Washington, DC. Garibaldi (1984) surveyed the black male students in the New Orleans Public Schools and found that 90 percent of them were interested in going to college. A similar survey of their teachers found that 60 percent of their teachers felt their black male students were not college material. What kind of mentoring does one suspect went on in the New Orleans Public Schools?

However, Xavier University takes some of the very same students (average combined SAT score in the 800 range), puts them in a rigorous prematriculation course and 55 percent of the undergraduate students are majoring in a science field—the highest percentage of any liberal arts college in the United States. Moreover, Xavier sends more of its black graduates to medical school than any other school except Howard University, and Xavier is a school with only 2,000 students—one-fifth the size of Howard.

This is an example of positive and negative mentoring in the same city. What does this suggest for the research study of mentoring? If similar situations were studied, quantitatively and qualitatively, we would have a more definitive definition of negative mentoring and a clearer perspective of its comparisions to positive mentoring.

The foregoing examples of negative mentoring illustrate that this phenomenon can have a profound effect on minority and disadvantaged students' career aspirations and academic achievement. But a signficant point is that negative mentoring can be exercised not with any negative intent but, indeed, with the students' "best" interest at heart.

DIMINISHING THE RESULTS

What can be done to diminish the results of negative mentoring, not in general, but particularly in those controlled situations where it is expected that a preponderant amount of negative mentoring occurs: in graduate study, in postgraduate study, and in entry into academe?

First we must study the phenomenon to gather the hard data that may corroborate the subjective impressions that are and have been obtained. The model for such a study could be partly based on a questionnaire administered by Braun (1990) to a stratified random sample of 300 lawyers in New York City. The questionnaire included the following question:

> At some time you may have had a relationship which looked like it had the potential of becoming a mentoring relationship, but in fact did not work out in this way. Perhaps the other person was a teacher or boss who at first seemed to have a great deal to offer, but the promise never was realized; the relationship did not develop or developed in a hurtful way.

In answer to this question, signficantly 20 percent of the men and 13 percent of the women reported a negative mentoring experience. Although none of the respondents was African American, this type of question issued to graduate students, postgraduate students, and new Ph.D.s entering the academy can elicit the existence and identification of subtle barriers of mentoring that may impede minority individuals' progress while ostensibly "aiding" minorities. Further examination of the phenomenon of negative mentoring will yield critcal data on this seldom studied experience.

ENHANCING THE OUTCOMES

Clearly, a more systematic study of the negative mentoring phenomenon needs to be done. That propsosition is reinforced by Braun's (1990) observation of, "Only a very few studies have even noted any negative aspects of the mentoring relationships." From my observation there have been even less studies of the relationship of the nonminority mentor with the racial minority protégé when the mentor does things ostensibly in the best interests of the mentee that result in diminishing the mentee's goals and lowering the mentee's expectations; indeed, sometimes leading to the mentee's career failure.

How can such behavior, which is, admittedly, mostly unconscious, be brought to consciousness and turned into a positive attribute? First, as stated earlier, it requires systematic study and documentation to determine its pervasiveness and generality. We must emphasize that not all minorities experience negative mentoring, but anecdotal experience suggests that a substantial number do. For example, Laura Rendón (1996) Professor of Education at Arizona State University, relates the experience of one atypical minority student:

> As a Puerto Rican woman from the upper-class sector of the island, I was born into privilege.... These are privileges I still hold because of my economic background, because of my social status as an intellectual, and because of the positive empowerment I felt throughout my youth and adolescence. These, however, are privileges that not many Latinos and Latinas hold.

So understanding that minorities have class and wealth differences should not lead us to approaching the mentoring experience as though minorities were just one homogeneous group. On the other hand, more typical is the experience of the African-American student also reported by Rendón:

> I find that [as a] freshman coming to college right out of high school, there's no sense of direction. There's no one here, as there wasn't [anyone here] for me. [For example, there's no one to sit down with and say to], "Well, I'm not certain what curriculum I want, what classes I need to be taking, how should I go about registration..."

Many students such as this working-class, first-generation college student has no older generation models to fall back on. This student badly needs mentoring, particularly in the counseling form, and a negative mentoring experience can distort the student's goals irreparably.

Thus, in addition to the systematic study of the mentoring relationship, there should be, within the proper context, structured training and workshops for mentors. It should not be assumed that because a person is willing, affable, and is of goodwill, that such a person will make a good mentor. A systematic study of mentoring programs, particularly with minority students, would aid in the construction of a good mentoring program that avoids the pitfalls of a negative mentoring experience.

Particularly in graduate school, where students are ordinarily not thought to need mentors, minority students especially need firm direction, encouragement to try new ideas and praise for their successes. Minority graduate students are ordinarily steered away from non-mainstream subjects and discouraged from exploring minority-related topics, which as a result often tends to stifle their creativity. Young professors also require positive mentoring and an older, senior professor to talk with. This is especially true for minority academics who may oftentimes be the only one in their department and, in their isolation, may take on too many committee assignments or may not know what is fully required to get tenure. These are situations where a negative mentoring experience can be professionally fatal.

A positive mentor is best described by Anderson and Ramey (1990) as one who: "challenges me, is objective, follows through, celebrates my successes, inspires loyalty, reassures me, is optimistic, is courageous, is perceptive, and empowers me." Moreover, mentors who are positive need not be of the same sex or race of the protégé (Braun, 1990).

The following is a list of programs oriented for undergraduate and graduate minority students that include mentoring components and that can serve as models of what a positive mentoring experience can be.

- NACME (National Action Council for Minorities in Engineering)
- MESA (Minority Engineering and Science Associates)
- FEF (Florida Endowment Fund)
- GEM (National Consortium for Graduate Degrees for Minorities in Engineering and Science, Inc.)
- GEM Consortium's National Institute on Mentoring
- Nissan-HBCU (Historically Black Colleges and Universities) Summer Institute

These are some of the minority higher education programs that have mentoring components. They can serve as models of what positive mentoring is about. Some of these programs are national in scope and others are regional. But they all strive to transform the educational experience in ways that Rendón (1996) so eloquently describes:

Our task is to create a new consciousness in the academic borderlands—one that heals, one that connects diverse cultures, languages, realities, and ways of knowing.

REFERENCES

Allen, W.R., Epps, E.G., & Hanniff, N.C. (Eds.) (1991). *College in black and white: African American students in predominantly white and historically black public universities.* Albany: State University of New York Press.

Anderson, R.T., & Ramey, P. (1990). Women in higher education: Development through administrative mentoring. In L.B. Welch (Ed.), *Women in higher education: Changes and challenges.* New York: Praeger Publishers.

Bowen, R., & Muller, G. (Eds). (1996). *Achieving administrative diversity.* San Francisco: Jossey-Bass Publishers.

Braun, R. (1990). The downside of mentoring. In L.B. Welch (Ed.), *Women in higher education: Changes and challenges.* New York: Praeger Publishers.

Crain, R.L. (1976). Why academic research fails to be useful. In F. Levinson & B. Wright (Eds.), *School desegregation: Shadow and substance.* Chicago: University of Chicago Press.

DeVos, G.A. (1973). *Socialization for achievement.* Berkeley: University of California Press.

Galbraith, M.W., & Cohen, N.H. (Eds.). (1995). *Mentoring: New strategies and challenges.* San Francisco: Jossey Bass Publishers.

Garibaldi, A. (1984). *Black colleges and universities: Challenges for the future.* New York: Praeger.

Herzog, R.A. (1982). High school senior occupational plans and values: Trends in sex differences 1976 through 1980. *Sociology of Education, 55* (1).

Rendón, L.I. (1996, November-December). Life on the border. *About Campus.*

Rist, R. (1980). Student social class and teacher expectations: The self-fulfilling prophecy in ghetto education. *Harvard Educational Review.*

Rosenfeld, R.A. (1980). Race and sex difference sin career dynamics. *American Sociological Review, 45.*

Rosenthal, R., & Jacobson, L.F. (1968). *Pygmalion in the classroom: Teacher expectations and pupils intellectual development.* New York: Holt, Rinehart and Winston.

Rubin, L.B. (1976). White against white: School desegregation and the revolt of middle America. In F. Levinson & B. Wright, *School desegregation: Shadow and substance.* Chicago: University of Chicago Press.

Sells, L.W. (1976). The mathematics filter and the education of women and minorities. Unpublished.

Willie, C.V. (1976). Racial balance or quality education? In F. Levinson & B. Wright (Eds.), *School desegregation: Shadow and substance.* Chicago: University of Chicago Press.

AN ANALYSIS OF FIVE PITFALLS OF TRADITIONAL MENTORING FOR PEOPLE ON THE MARGINS IN HIGHER EDUCATION

Theresa McCormick

Mentoring programs for people of color and white women need to become a basic part of university policy in support of equity and diversity on campuses across the country. Today, we live in a multicultural society which is reflected in highly diverse schools. Higher education will become even more diverse given the rapidly increasing proportion of minority students to majority students. By the twenty-first century over 30 percent of students in public schools will be minority; however, the teaching force of color (about 12.5% now) is dwindling. For example, the percentage of black teachers has gone from 7.8 percent to 6.9 percent since 1983, and Hispanics and other people of color make up only 3.5 percent of public school teachers (Haberman, 1988, p. 39). At the university level the picture is even more out of alignment in that few men and women of color are entering teacher education programs.

Diversity in Higher Education, Volume 1, pages 187-202.
Copyright © 1997 by JAI Press Inc.
All rights of reproduction in any form reserved.
ISBN: 0-7623-0086-8

The fact that there are many more white males in higher level administrative and faculty positions in academe results in mentor pools being almost exclusively white and male (Matczynski & Comer, 1991). Due to this scarcity of diverse faculty in higher education and in the mentoring pipeline, the urgency to increase efforts to hire and retain men and women of color and white women on university campuses has been in the forefront of various reports on higher education in the last decade (Tack & Patitu, 1992; Grund, 1989a; Monaghan, 1989; Gray, 1988).

Even though there has been steady growth since 1960 in the numbers of women of all ethnic groups enrolling in postsecondary education and earning baccalaureate and advanced degrees (OECD, 1986; Grund, 1989b), there has not been a comparable increase in numbers of women faculty members in universities and colleges. For example, data indicate that full-time female faculty in higher education institutions held only 27.6 percent of all faculty positions.

Of the total number of full-time faculty positions (470,673) held by men and women in all ranks in 1983, 112,267 (24.5%) were held by white women; 8,911 (1.8%) by black women; 2,184 (.5%) by Hispanic women; 3,187 (0.7%) by Asian women; and 355 (.07%) by Native American women (*Digest of Education Statistics*, 1988). Although many universities have an identified goal of increasing the number of white women and faculty of color, the data cited indicates the lack of progress toward that goal and the severity of the problem. A recent report on my university showed that it has made a net gain of *one* African-American faculty member over the last five years, in spite of all the touting of diversity and efforts to increase faculty of color (Bruner, 1996). This paper addresses the problems sketched above by providing a theoretical framework for their analysis, by exploring mentoring as a promising tool to help solve them, and by advocating the initiation of mentoring of men and women of color and white women as a policy in school systems and universities.

The political and judicial events and decisions at the federal level during the last 10 years have fostered a climate that denies the multicultural reality of U.S. society. Provincial, Eurocentric perspectives in universities and colleges reflect this denial of diversity. All students need to see themselves reflected in the schools they attend. This reflection should occur not only in the curriculum, but also in the faculty who deliver the courses. At present, the curriculum and faculty only provide "good" reflections for Anglo students.

Attracting, hiring, and retaining more teachers of color would obviously add much needed diversity to our colleges and universities. Also, a more diverse faculty is urgently needed to provide role models for students of color to counter the resurgence of racism on campuses and to make programs truly reflective of multicultural realities in society.

BENEFITS, PITFALLS, AND CHALLENGES OF MENTORING

At a large predominantly white male university in the Midwest, in 1988, Dr. Jacqueline Flemming, an African American professor from Barnard College and author of *Blacks in College* (1984), emphasized the importance of role models and mentors for the success of both minority students and faculty. Appropriate role models for students of color can positively affect their actual experiences in school and help overcome the isolation and vulnerability that so many of them face (Winkler, 1988). Having someone to emulate is especially needed by female students of color, since the contributions, values, experiences, and perspectives of their ancestors are practically invisible in the standard curriculum. Minority and women role models serve as "flesh and blood" reminders that minorities and women do make a difference.

THE BEST CONCEPTS OF MENTORING ROLES

One of the best concepts of mentoring is that it is a validation process of "real acceptance into that social collegial network which assists with promotion, advancement, and recognition..." (Elliott & Holmberg-Wright, 1983, p. 68). Discussing the functions of mentors, Simeone (1987, p. 101) says, "Perhaps most important, the sponsor helps to form within the protégé the sense of him or herself as a member of the profession, encouraging and fostering a self-image as a legitimate member of the community of scholars."

Whether for first-year teachers of color in public schools or for first-year minority or female students or faculty in teacher education, appropriate mentoring can play a decisive role in their retention, career development, and success. Barnes and Stiasny (1995) emphasize the overall importance of mentoring in the process of teacher education. Mentoring enhances a newcomer's likelihood of success in the system by providing necessary information (e.g., how to obtain resources), counseling and guidance, and by developing trust and personal relationships/friendships. A mentor specifically introduces and initiates the mentee into the demands and expectations of school life, shares knowledge, provides encouragement, and serves as a role model. Most importantly, a mentor helps to create within the newcomer a sense of identity as a member of the profession. Here is a sampler of mentor roles drawn from the literature (see Table 1) (Odell, 1990, p. 7).

Since "traditional" mentoring, the avenue for success of white males in academia, has largely excluded men and women of color and white women, different and varied approaches to mentoring must be developed and implemented by universities and colleges to give the marginalized or excluded ones access to information and to other types of assistance that will enable them to succeed.

Table 1. A Sampler of Mentor Roles Drawn from the Literature

Mentor Roles	*Reference*
Trusted Guide	Homer's *Odyssey*
Teacher, Sponsor, Host, Counselor Supporter, Guru, Advisor	Levinson, 1978
Teacher Coach, Trainer, Positive Role Model, Developer of Talent, Opener of Doors, Protector, Sponsor, Leader	Schein, 1978
Traditional Mentor, Supporitve Boss Organizational Sponsor, Professional Mentor, Patron, Invisible Godparent	Phillip-Jones, 1982
Guide, Supporter, Challenger	Daloz, 1983
Confidant (in Addition to Schein's)	Gehrke & Kay, 1984
Master Teacher, Teacher Adviser Specialist, Researcher-linker	Bird, 1985
Colleague Teacher, Helping Teacher Peer Teacher, Support Teacher	Borko, 1986
Teacher, Sponsor, Encourager Counselor, Befriender	Anderson & Shannon, 1988

Source: Odell (1990).

SOME "PITFALLS" OF MENTORING

There is extensive documentation of numerous pitfalls of mentoring for men and women of color and white women to be wary of. Many of the acknowledged problems of mentoring for diverse men and women stem from the fact that it has been the main tool used by "the white male club" to maintain its nearly exclusive position of power in education. Carter (1982, p. 4) says, "While much of the literature stresses the importance of the mentor-protégé relationship..., it must also be noted that success in mentor-protégé relationships seems assured only when both partners in the relationships are white males."

My position is that the present model of "traditional" white male mentoring *does* contain pitfalls for men and women of color and white women, but it can and must be transformed to be equally functional for our academic advancement since the evidence so clearly indicates the necessity of having a mentor for success in academia or in any large organization. To transform mentoring to a model that is serviceable for diverse people in universities and colleges, the pitfalls must be delineated and solutions developed to overcome them.

Following is a summary of five key pitfalls of "traditional" mentoring that are identified in the literature. Later, an analysis of them is presented in the context of society and education institutions, and finally, some implications for higher education policy are offered.

Pitfall #1: "Traditional" mentoring promotes competition and focuses too much on personal ambition of the individual and promotes elitism and exclusion (Table 2) (Collins, 1982; Wilkerson, 1984)

Carter (1982, p. 9) notes, because of negative connotations about mentoring, many minority women who have "made it" on their own, "... deny the existence or the necessity of the mentor-protégé relationship as a factor in success... these successful minority and women professionals perceive themselves as elites and do not realize that they are on the periphery of the system."

Pitfall #2: Scarcity of senior level, appropriate mentors for minority women (Table 2).

Table 2. Five Pitfalls of "Traditional" Mentoring for Minorities and Women

Pitfall	*Description*
# 1	"Traditional" mentoring promotes competition and focuses too much on personal ambition of the individual and thus promotes elistism and exclusion (Collins, 1982; Wilkerson, 1984).
# 2	Scarcity of senior level, appropriate mentors for minorities and women (Hall & Sandler; 1983; Collins, 1982; Carter, 1982).
# 3	"Traditional" mentoring promotes and maintains the status quo by socializing protégés into the "rules of the game" and many of the "rules" one must learn in order to be in the "inner circle" are discriminatory against women and minorities (Collins, 1982).
# 4	Cross-race and cross-gender mentor-protégé relationships have not with success due to personal and organizational barriers (Kram, 1985; Zey, 1984; Collins, 1983).
# 5	"Traditional" mentor-protégé relationships promote dependency and subordination of minorities and women, and it is difficult for the protégé to move to the position of peer/colleague (Kram, 1985; Nichols & Golden, 1982; Collins, 1983; Carter, 1982; Hall & Sandler, 1983).

This is linked with pitfall #1 (competition) (Collins, 1982). Minority women must compete for a mentor from the existing pool of senior faculty who are usually white and/or male and this competition for a limited resource can create conflict by undermining cooperation and solidarity of the group (Hall & Sandler, 1983; Collins, 1982). Carter (1982, p. 5) observes, "...the literature documents a glaring reality concerning the lack of successful mentor relationships for young minority persons and women."

Pitfall #3: "Traditional" mentoring promotes and maintains the status quo by socializing protégés into the "rules of the game" and many of the "rules" one must learn in order to be in the "inner circle" are discriminatory against faculty women and people of color (Table 2)

Roger Collins (1982, p. 7) equates learning and accepting the "rules of the game" with "...colonialist strategies that require the colonized to abdicate their identity, their history, and their commitment to oppressed compatriots in order to receive colonial favors apparently crucial to their survival." Part of the inherent conflict for a woman of color being mentored into the "inner circle" of academia concerns the inculcation of an intellectual legacy that may be incompatible with her own paradigms of knowledge and the world (Collins, 1982).

Pitfall #4: Cross-race/cross-gender mentor-mentee relationships have not met with success due to personal and organizational barriers (Table 2) (Kram, 1985; Zey, 1984; Collins, 1982; Collins, 1983).

In studies on mentoring in the corporate world both Zey (1984) and Kram (1985) mention "comfort-discomfort" factors involved in cross-gender mentoring. These concerns are due in part to the lack of socialization of both sexes to interact with each other as peers in close work relationships. Discussing her work on cross-gender mentoring, Kram (1985, p. 64) says, "Concerns about increasing intimacy and concerns about the public image of the relationship caused both individuals to avoid interaction that had the potential to provide a wide range of career and psychosocial functions;... Similar patterns are likely to exist in interracial relationships as well." It is important to note that the research does *not* negate the value of mentoring (whether cross-gender or cross-race), but rather it indicates, as Carter (1982, p. 8) says, "...that the mentor-protégé relationship functions most effectively when the mentor and the protégé share common values, attitudes, goals, and perceptions of the world."

Pitfall #5: "Traditional" mentor-mentee relationships promote dependency and subordination of women of color, and it is difficult for the mentee to move to the position of peer/colleague (Table 2).

This pitfall is linked with the charges that mentoring is paternalistic/ maternalistic, colonalist, and elitist (Kram, 1985; Collins, 1983; Carter, 1982;

Hall & Sandler, 1983; Collins, 1982). Nichols and Golden (1982, p. 9) strongly state, "Mentors can inhibit independence and professional autonomy. And the evidence indicates that even more than others, minority persons and women are expected,...to remain subordinate...and dependent."

BEYOND THE WHITE MALE CLUB

The ongoing societal and institutional conditions that support the continued low number of faculty men and women of color in universities and the slow trickle of teachers of color in the pipeline may be examined by using an analytic tool developed by Robert Terry (1974, 1986) which focuses on four interdependent aspects of organizations: mission/culture, power, structure, and resources (Table 3). In the process of this analysis the five identified pitfalls of mentoring for men and women of color and white women in education will be critiqued as symptoms of a larger organizational dysfunction of "the white male club." Hopefully, the critique will help clarify the pitfalls so that a more functional model of mentoring for minorities and women can be implemented.

Table 3. Robert Terry's Model for Analyzing Organizations
Mission/Culture

Resources	*Structure* *Power* *Four Interdependent Aspects of the Model*
1. Mission	The direction "toward which" the organization moves Implicit vision, dreams, hopes, meaing, aspiration of people in organizations
2. Power	The enegry "by which" the organization makes and keeps decisions over time Entails commitment and will Access to resources Mobilization of resources to achieve goals
3. Structure	The form "through which" everyday affairs are carried out Institutionalized patterns Practices and policies—define roles Allot rewards and punishments
4. Resources	The materials "from which" energy flows to make things happen in an organization Concrete materials Time, space, and status Wealth and income derived from the organization

In workshop settings (1986), as well as in his writing (1974), Terry, consultant on racism, sexism, and institutional change, has conceptualized and illustrated his four-point model for analyzing problems in an organization as hierarchical beginning with its mission/culture, then the power considerations, followed by its structure, and concluding with resources of the organization (Table 3). The four concepts are interrelated, but mission/culture are considered to be primary since their omnipresence fundamentally affects all other aspects of an organization.

Mission as conceptualized by Terry (1974, 1986) is a concept that describes the direction "toward which" the organization moves. The mission incorporates the meaning, vision, dreams, hope, aspirations that are implicit in the human endeavors of the organization. *Culture* is embedded in the mission of the organization. The culture of the group includes the ethos, values and value orientations, beliefs, and standards assumed to be shared by group members.

Power in Terry's model embodies the energy "by which" the organization makes and keeps decisions over time and entails commitment and will. In addition to having access to resources, holders of power can mobilize resources to accomplish their goals.

Structure is the form "through which" the everyday business of the organization is carried out. Structure is provided by institutionalized patterns, practices, and policies that define roles and allot rewards and punishments. It is embodied by "rules of the game," regulations, and procedures that are patterned over time.

Resources are the "materials" "from which" energy flows to make things happen in an organization. Resources encompass concrete materials as well as time, space, status and of course, wealth and income derived from the organization. Holders of power have access to organizational resources, control their distribution pattern, and can use them to achieve their goals.

ANALYSES AND STRATEGIES

To analyze institutional problems, such as the low number of men and women of color on the faculty in universities or the pitfalls of mentoring for them, Terry (1974, 1986) believes that one must first determine the nature of the problem. Is it a mission/culture problem, a power, structure, or a resources problem? In his consulting experience with varied organizations across the United States, he says generally when people agree on the nature of a problem (e.g., power), it actually is the one above it in the hierarchy (i.e., mission) and they tend to try to solve the problem at the level below (i.e., structure). Another issue that he has identified is that some groups stay at the lower levels of analysis (structure and resources) in their effort to bring about change and never deal with power and mission issues in the organization.

The cultural standards of "the white male club" engender misplacement of the problem, and to avoid changing, the club blames the victim for its problems (Terry, 1974). He also observes, "To examine racism and sexism one must look beyond facts and statistics, ... Racism and sexism in America are not problems simply to be listed alongside other problems. They are part of the club's foundation" (p. 18).

ANALYSIS OF PITFALL #1

"The white male club" in highly technical, Western cultures values and "rewards individual initiative, encourages competition, supports dispassionate, objective, and analytic thinking,..." according to Terry (1974, p. 18). Competitive value orientations permeate large research universities in the United States; however, many women of color place more value on communal sharing, collaboration, and openness and feel that cutthroat competitiveness is debilitating and dysfunctional in their careers.

Pitfall #1 is analyzed as being a problem related to the university's culture with its white male values, value orientations, beliefs, and standards that are erroneously assumed to be shared by all who are in the institution.

Transforming strategies to correct pitfall #1 include the following ideas which address the prime level of Terry's analytic tool, that is, mission and culture. The culture and mission of white male-dominated universities need to be humanized and redirected, not totally eliminated. Terry (1974, p. 18) says, "Bringing together the strengths of both sets of values [male/female, minority/majority]—a process of cultural synergism—is absolutely essential if any alternative to the club is to emerge."

Since the present cultural values of universities, based on an "either/or," exclusionary mind-set, feed competition and conflict, a "both/and" approach must be worked toward in order to make mentoring workable for diverse faculty on the margins. If the mission of education is to have meaning for *all* its constituents, leaders must redirect its course toward community and connectedness. As Terry (1986) explains, persons in communities exist; "we are, therefore, I am."

Another orientation for redirecting the negative effect of "traditional" mentoring for men and women of color and white women is found in Gehrke's (1988) exposition of Martin Buber's I-Thou and I-It philosophy of primal relationships. The common paradigm in use is the I-It relationship which is manipulative and pragmatic, whereas, the I-Thou relationship abandons objectifying and manipulating the other for an authentic confirmation of each one's existence. She says, "...if we wish to encourage mentoring roles among a variety of helping roles, we should attend to the context and conditions for achieving more of an I-Thou relationship" (p. 44).

Still another approach to reducing the competition of mentoring is illustrated by Wilkerson's (1984) conception, "Lifting as We Climb." The idea of helping others as one advances has roots in African-American's struggles for freedom from slavery. Wilkerson (1984, p. 61) says, "It ["Lifting as We Climb"] is still pertinent to women of color...for it acknowledges their need and desire for personal advancement while connecting them with a collective effort to improve the quality of life for many."

ANALYSIS OF PITFALL #2

Conflict arises because of scarcity of a desired resource (appropriate senior level mentors) for minority women faculty. As we have observed in Terry's analytic model, holders of power can command resources to achieve their goals, and white males control a disproportionate share of resources. Roger Collins (1982) pinpoints the problem by saying,

> For women and minority persons, the scarcest and most valuable resource is often the availability of a mentor within a particular organization...it is likely that mentors select protégés on the basis of their similarities...this tendency diminishes opportunities for selection for those members of groups that have been traditionally excluded from the organization (p. 4).

Pitfall #2 is analyzed initially as a resource problem, as inferred from the previous discussion. Unequal distribution of resources also indicates that power and structure problems exist. Since this pitfall is so closely tied to pitfall #1 (competition), a mission/culture problem is indicated also.

Transforming strategies to address the resource, mission/culture, power, and structure problems embedded in pitfall #2 include augmenting a formal mentoring program with informal systems such as networking, peer pals, "sister" systems, paper mentors (for example, written materials on "learning the ropes" of the tenure and promotion process) to provide adequate resources for the mentoring needs of women of color faculty.

Likewise, it is suggested that structural rewards, such as faculty released time, merit raises, credit toward tenure and promotion, or at least, public recognition be built into the system for mentors. Since the few minority faculty women who do hold senior positions are usually "stretched thin" serving as the "token minority" on numerous committees, they especially need tone awarded released time for mentoring junior faculty.

ANALYSIS OF PITFALL #3

Pressure to conform to the "rules of the game" and to role expectations is as powerful in white male-dominated universities as in any other institution.

Adherence to "rules of the game" is a price to be paid by the mentee in "traditional" mentoring relationships. Roger Collins (1982, p. 6) says, "...to the extent that certain of these rules have served to undermine the status of women and minority faculty, they are likely to reject them."

Pitfall #3 is analyzed primarily as a structure problem, since it is the rules, regulations, and policies defining the structure of universities that are problematic in the context of this pitfall. Indications of a mission/culture problem are also implied because of the ways in which "the club" deals with dissenters [by pacification or elimination] (Terry, 1974).

Transforming strategies to help alleviate the structure admission/culture problems related to pitfall #3 address the need to overcome the rejection of having a single "traditional" mentor by minority women who feel that they will be "colonized" or that they will sacrifice their personal integrity. Two options that Carter (1982, p. 11) mentions are initiating multiple mentor relationships with different senior faculty and establishing symbolic mentorships wherein the junior faculty "have the opportunity to observe academia function without locking themselves into a relationship which they deem condescending or colonizing." Fisher (1994) acknowledges the value of having a program where multiple roles and varied types of mentoring are part of the organization.

Structural problems associated with pitfall #3 also can be addressed by examining the school policy concerning allotment of "rewards and punishments" associated with institutional practice. Flemming (1984) says that providing a mentoring program is a structural issue since time given to a junior faculty member detracts from the senior faculty member's research time. Therefore, she advocates that mentoring be legitimized as a part of the reward system for faculty. When a reward system is built into institutional policy, then accountability must be addressed by conducting a systematic assessment of mentors and the success rate of their mentees.

Another policy that would assure appropriate mentoring for minority and women faculty becomes a reality in the structure of the school is to make the track record of hiring and retaining minority and women faculty a component in the evaluation of college deans, department chairs, and school administrators. Restructuring of rules, procedures for job searches, and other practices can be hastened by meshing school-wide mentoring efforts and accountability measures with the affirmative action program and administrative efforts.

ANALYSIS OF PITFALL #4

A high rate of success in mentoring relationships occurs when both mentor and mentee are white males, while the failure rate is higher with women and minorities. Typically, "The club" blames the victim for not meeting the required standards.

Pitfall #4 is analyzed as a mission/culture problem since the lack of success of "traditional" mentoring for minority females has been attributed to the fact that the culture embedded in the mission of white male-dominated universities takes care of its own and only passively includes those who are "different." Also, there are structural and power problems associated with this pitfall because of the institutional practices that have evolved from the values of "the club."

A *transforming strategy* related to solving the culture difference problem is to start with same sex/same race mentors/mentees if possible (Hall & Sandler, 1983). Then later, the initial mentor helps the mentee find another mentor with a different background and different strengths. It is recommended that mentees seek both types of mentors—those of their own race and sex *and* those who command the power in the area that they are trying to enter.

To increase the "comfort" level of interaction between cross-sex/cross-race mentoring relationships, institutions must develop more sex- and race-neutral activities and sites where social contact occurs naturally and comfortably. Many corporations are developing a more familial environment by organizing special interest clubs (exercise, computers) where all people can mix, in contrast to exclusive male clubs of the past (Zey, 1984).

Structural problems related to pitfall #4 may be addressed by attending to an idea provided by Kram (1985). She indicates that the reward system must be changed from one that values only bottom-line results to one that values how well senior members help develop their juniors. She also suggests evaluating the task design of the workplace. If it is found to be isolating and competitive then change the design of work tasks to encourage relationship building and interaction between people.

ANALYSIS OF PITFALL #5

Egalitarian principles do not prevail over the process of gaining access to advancement in education. Access to professional authority and success in academia is problematic for minority women because of the difficulty of progressing from the dependency of mentee status to the level of full colleague. In *Passages*, Gail Sheehy (1974) observes that mentees must let go of the mentor in order to grow to their full potential. When it is time for the relationship to end, Nancy Collins (1983, p. 34) warns, "Women must guard against any impulse not to pass men."

Pitfall #5 is analyzed initially as a structural problem because dependency of subordinates is a significant part of the foundation of institutional racism and sexism. Practices in modern-day institutions reflect the persistence of attitudinal remnants of early common law wherein women and children were legally controlled by a paternal power and where the legal status of slaves was equated with that of women and children. The latter points are made in Gunnar Myrdal's classic work on racial discrimination, *An American Dilemma* (1944). "Fallout" from this long-term structural problem includes "club members'" stereotypic views that minorities and women are intellectually inferior and not well prepared for independent leadership and scholarship. Thus, they are admitted on a token basis, winnowed out by the "objective" rules and regulations, and then blamed for not measuring up to the institution's standards. From the preceding discussion, it is clear that power and mission/culture problems underlie the structural issues.

Transforming strategies for addressing problems associated with pitfall #5 overlap considerably with those discussed for pitfalls #1, 2, 3, and 4. In brief, the key strategy is to develop alternative approaches to mentoring for minorities and women so that they are not solely dependent on white male senior educators. While the alternative approaches discussed earlier may not tap directly into the "inner circle" of powerholders, they can have a positive influence and be a beneficial source of information and support. An additional reminder to help the mentee move through the separation phase with a mentor and on to a position of peer and colleague is to remember the cyclical nature of mentoring relationships (Collins, 1983). Developing mentoring relationships needs to be perceived as a *process* that is ongoing, with present needs being met and not perceived as a product that is cast in concrete once a relationship is established.

FIVE PRINCIPLES FOR HIGHER EDUCATION POLICYMAKERS TO CONSIDER

The multifaceted approach that I propose for universities and schools to consider for implementation is guided by the following five principles:

1. Initiators of mentoring programs for men and women of color and white women attend to all four points of Terry's (1974, 1986) analytic model before implementing a formal program.
2. A successful mentoring program for diverse faculty is sensitive to their unique needs and adapted to their different backgrounds and aspirations.
3. The program is formal with an explicit policy that provides legitimation from the highest officials. It is augmented and supported by informal networks, alternative support systems, varied resources, and rewards for mentors.
4. The program is conceptualized as an ongoing, and evolving *process*, not as a narrow, bottom-line product to satisfy affirmative action.

5. The mentoring program is conceptualized as a vehicle for collective sharing, cooperation, and community connectedness. The supporting institutional ethos is the development of human resources and quality of the educational climate.

CONCLUSION

Increasing the numbers of men and women of color and white women faculty on university campuses is urgently needed to provide an accurate reflection of societal diversity and to provide appropriate role models for a diverse student body. Changes proposed in this paper for transforming mentoring not only would help men and women of color and white women, but they also would benefit education in general by providing a model of active inclusion to be used with other underrepresented groups, such as gays and lesbians, and people with disabilities.

By providing a mentoring program that is functional for men and women of color and white women, universities and colleges would be taking an active nonracist, nonsexist stance. As Robert Terry (1986) says, authentic pluralistic institutions practice *active inclusion* that is based on "a new consciousness of color and gender." I agree with Terry when he notes that a passive, nonracist, nonsexist stance by a person or an institution is an illusion.

People who now reside on the margins of academia must be brought to the inner circle of university life if transformation is to occur. Universities, like other institutions, are self-perpetuating and not notably responsive to outside pressure for change. More than admitting men and women of color and white women faculty must be done in order for universities to be active in the cause for equity. Programs to retain diverse faculty as fully engaged insiders in the system must be implemented, and I believe that appropriate mentoring is a critical element in achieving that goal.

ACKNOWLEDGMENT

Slightly different versions of this paper have appeared in McCormick (1992) and McCormick and Titus (1990).

REFERENCES

Anderson, E.M., & Shannon, S.L. (1988). Toward a conceptualization of mentoring. *Journal of Teacher Education, 39* (1), 38-42.

Barnes, T., & Stiasny M. (1995). *Mentoring: Making it work–Partnership in teacher education.* Southampton, England: The Bassett Press.

Bird, T. (1985). From teacher to leader: Training and support for mentor teachers, master teachers, and teacher advisors. Unpublished manuscript. San Francisco: The Far West Regional Laboratory.

Borko, H. (1986). Clinical teacher education: The induction years. In J.V. Hoffman & S.A. Edwards (Eds.), *Reality and reform in clinical teacher education* (pp. 45-63). New York: Random House.

Bruner, J. (1996, November 19). ISU still has retention, recruitment problems. *The Daily Tribune's University Times*, p. U4.

Carter, H. (1982, March). Making it in academia: Gurus can get you there? Paper presented at the annual meeting of the American Educational Research Association in New York (ERIC Document Reproduction No. ED 235 758).

Collins, N. (1983). *Professional women and their mentors–A practical guide to mentoring for the woman who wants to get ahead.* Englewood Cliffs, NJ: Prentice Hall.

Collins, R. (1982, March). Colonialism on campus: A critique of mentoring to achieve equity in higher education. Paper presented at the annual meeting of the American Educational Research Association in New York (ERIC Document Reproduction No. ED 235 760).

Daloz, L.A. (1983). Mentors: Teachers who make a difference. *Change, 15* (6), 24-27.

Elliott, P., & Holmberg-Wright, K. (1983, Fall). Jewel tea and academe. *Action in Teacher Education, 5* (3), 67-72.

Fisher, B. (1994). *Mentoring*. London, England: Library Association Publishing Ltd.

Flemming, J. (1984). *Blacks in college*. San Francisco, CA: Jossey-Bass.

Gehrke, N. (1988, January-February). On preserving the essence of mentoring as one form of teacher leadership. *Journal of Teacher Education, 34* (1), 43-45.

Gehrke, N., & Kay, R.S. (1984). Socialization of beginning teachers through mentor-protégé relationships. *Journal of Teaching Education, 35* (3), 12-14.

Gray, M. (1988, December 14). Bush must act quickly on civil rights issues if he is to create a "kinder, gentler nation." *The Chronicle of Higher Education, 35* (16), B1-B2.

Grund, N. (Ed.). (1989a, March 13). Full-time faculty in higher education by race/ethnicity—Minority faculty initiatives cited. *Higher Education & National Affairs, 38* (5), 1, 4.

Grund, N. (Ed.). (1989b, May 8). Facts in brief—Women earning more doctorates. *Higher Education & National Affairs, 38* (9), 3.

Haberman, M. (1988, July-August). Proposals for recruiting minority teachers: Promising practices and attractive detours. *Journal of Teacher Education, 39*(4), 38-44.

Hall, R., & Sandler, B. (1983). Academic mentoring for women students and faculty: A new look at an old way to get ahead. Washington, DC: Association of American Colleges. (ERIC Document Reproduction No. ED 240 891).

Kram, K. (1985). *Mentoring at work/Developmental relationships in organizational life*. Glenview, IL: Scott, Foresman.

Levinson, D. (1978). *The seasons of a man's life*. New York: Knopf.

Matczynski, J., & Comer, K. (1991). Mentoring women and minorities in higher education. *Viewpoints* (ERIC Document Reproduction No. ED 331 376).

McCormick, T., & Titus, P. (1990). *Mentoring International, 4* (2), 31-37

McCormick, T. (1992). *Proceedings of the second annual National Association of Multicultural Education conference*. Morristown, NJ: Silver Burdett Ginn.

Monaghan, P. (1989, March 8). Action, not just policy change, seen needed to improve minority scholars' opportunities. *The Chronicle of Higher Education, 35* (26), A18.

Myrdal, G. (1944). *An American dilemma*. New York: Harper and Row.

National Center for Education Statistics (1988). *Digest of education statistics 1988*. Washington, DC: U.S. Department of Education.

Nichols, I., & Golden, P. (1982, March). The middle way: Patron as guide. Paper presented at the annual meeting of the American Educational Research Association in New York (ERIC Document Reproduction No. ED 235 759).

Odell, S.J. (1990). Support for teachers. In T.M. Bey & C.T. Holmes (Eds.), *Mentoring: Developing successful new teachers* (pp. 3-23). Reston, VA: Association of Teacher Educators.

Organization for Economic Co-Operation and Development (OECD) (1986). *Girls and women in education–A cross-national study of sex inequalities in upbringing and in schools and colleges.* Paris: Author.

Phillips-Jones, L. (1982). *Mentors and proteges.* New York: Arbor House.

Schein, E. (1978). *Career Dynamics: Matching individual and organizational needs.* Reading, MA: Addison-Wesley.

Sheehy, G. (1974). *Passages: Predictable crises in adult life.* New York: E.P. Dutton.

Simeone, A. (1987). *Academic women–Working towards equality.* South Hadley, MA: Bergin and Garvey.

Tack, M.W., & Patitu, C.L. (1992). *Faculty job satisfaction: Women and minorities in peril.* ASHE-ERIC Higher Education Report No. 4. Washington DC: The George Washington University, School of Education and Human Development.

Terry, R. (1974, Spring). The white male club: Biology and power. *Civil Rights Digest, 6* (3), 66-77.

Terry, R. (1986, April 10). Leadership and civil rights. Paper presented at the Midwest Human Relations Conference in Minneapolis.

U.S. Supreme Court Bulletin. (1989, May 8). *City of Richmond v. Croson Co. 49* (36A), B 764-B860.

Wilkerson, M. (1984, March). Lifting as we climb: Networks for minority women. *Women in Higher Education Administration New Directions for Higher Education, 12* (45), 59-66.

Winkler, K. (1988, November 9). Minority students, professors tell of isolation, anger in graduate school. *The Chronicle of Higher Education, 35* (11), A15-A16.

Zey, M. (1984). *The mentor connection.* Homewood, IL: Dow Jones-Irwin.

www.ingramcontent.com/pod-product-compliance
Ingram Content Group UK Ltd.
Pitfield, Milton Keynes, MK11 3LW, UK
UKHW020843290926
15470UKWH00004B/1529